T0140632

Advances in Intelligent Systems and Computing

Volume 901

Series editor

Janusz Kacprzyk, Systems Research Institute, Polish Academy of Sciences, Warsaw, Poland
e-mail: kacprzyk@ibspan.waw.pl

The series "Advances in Intelligent Systems and Computing" contains publications on theory, applications, and design methods of Intelligent Systems and Intelligent Computing. Virtually all disciplines such as engineering, natural sciences, computer and information science, ICT, economics, business, e-commerce, environment, healthcare, life science are covered. The list of topics spans all the areas of modern intelligent systems and computing such as: computational intelligence, soft computing including neural networks, fuzzy systems, evolutionary computing and the fusion of these paradigms, social intelligence, ambient intelligence, computational neuroscience, artificial life, virtual worlds and society, cognitive science and systems, Perception and Vision, DNA and immune based systems, self-organizing and adaptive systems, e-Learning and teaching, human-centered and human-centric computing, recommender systems, intelligent control, robotics and mechatronics including human-machine teaming, knowledge-based paradigms, learning paradigms, machine ethics, intelligent data analysis, knowledge management, intelligent agents, intelligent decision making and support, intelligent network security, trust management, interactive entertainment, Web intelligence and multimedia.

The publications within "Advances in Intelligent Systems and Computing" are primarily proceedings of important conferences, symposia and congresses. They cover significant recent developments in the field, both of a foundational and applicable character. An important characteristic feature of the series is the short publication time and world-wide distribution. This permits a rapid and broad dissemination of research results.

More information about this series at http://www.springer.com/series/11156

Rafael Valencia-García · Gema Alcaraz-Mármol
Javier del Cioppo-Morstadt · Néstor Vera-Lucio
Martha Bucaram-Leverone
Editors

ICT for Agriculture and Environment

Second International Conference, CITAMA 2019, Guayaquil, Ecuador, January 22–25, 2019, Proceedings

Editors
Rafael Valencia-García
Facultad de Informática
Universidad de Murcia
Murcia, Spain

Gema Alcaraz-Mármol
Departamento de Filología Moderna
Universidad de Castilla–La Mancha
Toledo, Spain

Javier del Cioppo-Morstadt
Universidad Agraria del Ecuador
Guayaquil, Ecuador

Néstor Vera-Lucio
Universidad Agraria del Ecuador
Guayaquil, Ecuador

Martha Bucaram-Leverone
Universidad Agraria del Ecuador
Guayaquil, Ecuador

ISSN 2194-5357 ISSN 2194-5365 (electronic)
Advances in Intelligent Systems and Computing
ISBN 978-3-030-10727-7 ISBN 978-3-030-10728-4 (eBook)
https://doi.org/10.1007/978-3-030-10728-4

Library of Congress Control Number: 2018965453

This Springer imprint is published by the registered company Springer Nature Switzerland AG
The registered company address is: Gewerbestrasse 11, 6330 Cham, Switzerland

CITAMA 2019—Preface

Agronomy and environment represent a key element in the present-day society. These areas have great importance in Latin American countries, technological innovations in these areas representing a key strategy. Particularly, the application of ICTs to agronomy and environment is a current need and there are new and exciting innovations in this field, this issue becoming one of the most advanced sectors of digital technologies.

The 2nd International Conference on ICTs in Agronomy and Environment (CITAMA 2019) was held during January 22–25, 2019, in Guayaquil, Ecuador. The CITAMA series of conferences try to become an international framework and meeting point for professionals who are particularly devoted to research, development, and innovation, within the field of the application of ICTs to agronomy, the environment, and related areas. CITAMA 2019 was conceived as a knowledge exchange conference consisting of several plenary talks and contributions about current innovative technology. These proposals deal with the most important aspects and future prospects from an academic, innovative, and scientific perspective, within the field of applied ICTs in agronomy and environment. It aims to provide a platform for researchers to share their original work and practical development experiences.

We would like to express our gratitude to all the authors who submitted papers to CITAMA 2019, and our congratulations to those whose papers were accepted. There were 27 submissions this year. Each submission was reviewed by at least three Program Committee (PC) members. Only the papers with an average score of 1.0 were considered for final inclusion, and almost all accepted papers had positive reviews or at least one review with a score of 2 (accept) or higher. Finally, the PC decided to accept 14 full papers. We would also like to thank the Program Committee members, who agreed to review the manuscripts in a timely manner and provided valuable feedback to the authors.

January 2019

Rafael Valencia-García
Gema Alcaraz-Mármol
Javier del Cioppo-Morstadt
Néstor Vera-Lucio
Martha Bucaram-Leverone

Organization

Honor Committee

Martha Bucaram Leverone	Universidad Agraria del Ecuador, Ecuador
Javier del Cioppo-Morstadt	Universidad Agraria del Ecuador, Ecuador
Néstor Vera Lucio	Universidad Agraria del Ecuador, Ecuador
Tayron Martinez	Universidad Agraria del Ecuador, Ecuador

Organizing Committee

Rafael Valencia-García	Universidad de Murcia, Spain
Gema Alcaraz-Mármol	Universidad de Castilla–La Mancha, Spain
Javier del Cioppo-Morstadt	Universidad Agraria del Ecuador, Ecuador
Néstor Vera Lucio	Universidad Agraria del Ecuador, Ecuador
Martha Bucaram Leverone	Universidad Agraria del Ecuador, Ecuador

Program Committee

Jacobo Bucaram Ortiz	Universidad Agraria del Ecuador, Ecuador
Martha Bucaram Leverone	Universidad Agraria del Ecuador, Ecuador
Rina Bucaram Leverone	Universidad Agraria del Ecuador, Ecuador
Rafael Valencia-García	Universidad de Murcia, Spain
Ricardo Colomo-Palacios	Ostfold University College, Norway
Ghassan Beydoun	University of Technology Sydney, Australia
Antonio A. López-Lorca	University of Melbourne, Australia
José Antonio Miñarro-Giménez	Medical Graz University, Austria
Catalina Martínez-Costa	Medical Graz University, Austria

Chunguo Wu	Jilin University, China
Siti Hajar Othman	Universiti Teknologi Malaysia (UTM), Malaysia
Anatoly Gladun	V. M. Glushkov of National Academy Science, Ukraine
Giner Alor-Hernández	Instituto Tecnológico de Orizaba, Mexico
José Luis Ochoa	Universidad de Sonora, Mexico
Ana Muñoz	Universidad de Los Andes, Venezuela
Miguel Ángel Rodríguez-García	Universidad Rey Juan Carlos, Spain
Lucía Serrano-Luján	Universidad Rey Juan Carlos, Spain
Eugenio Martínez-Cámara	Universidad de Granada, Spain
Gema Alcaraz-Mármol	Universidad de Castilla–la Mancha, Spain
Francisco M. Fernandez-Periche	Universidad Antonio Nariño, Colombia
Ali Pazahr	Islamic Azad University—Ahvaz branch, Iran
Mónica Marrero	Delft University of Technology, Netherlands
Alejandro Rodríguez-González	Universidad Politécnica de Madrid, Spain
Carlos Cruz-Corona	Universidad de Granada, Spain
Dagoberto Catellanos-Nieves	Universidad de la Laguna, Spain
Juan Miguel Gómez-Berbís	Universidad Carlos III de Madrid, Spain
Jesualdo Tomás Fernández-Breis	Universidad de Murcia, Spain
Francisco García-Sánchez	Universidad de Murcia, Spain
Antonio Ruiz-Martínez	Universidad de Murcia, Spain
Manuel Quesada-Martínez	Universidad Miguel Hernández, Spain
Maria Pilar Salas-Zárate	Instituto Tecnológico de Orizaba, Mexico
Mario Andrés Paredes-Valverde	Instituto Tecnológico de Orizaba, Mexico
Luis Omar Colombo-Mendoza	Instituto Tecnológico de Orizaba, Mexico
Katty Lagos-Ortiz	Universidad Agraria del Ecuador, Ecuador
José Medina-Moreira	Universidad de Guayaquil, Ecuador
José María Álvarez-Rodríguez	Universidad Carlos III de Madrid, Spain
Thomas Moser	St. Pölten University of Applied Sciences, Austria
Lisbeth Rodriguez Mazahua	Instituto Tecnologico de Orizaba, Mexico
Jose Luis Sanchez Cervantes	Instituto Tecnologico de Orizaba, Mexico
Cristian Aaron Rodriguez E.	Instituto Tecnologico de Orizaba, Mexico
Humberto Marin Vega	Instituto Tecnologico de Orizaba, Mexico
María Teresa Martín-Valdivia	Universidad de Jaén, Spain
Miguel A. García-Cumbreras	Universidad de Jaén, Spain
Begoña Moros	Universidad de Murcia, Spain

Salud M. Jiménez Zafra	Universidad de Jaén, Spain
Arturo Montejo-Raez	Universidad de Jaén, Spain
M. Abirami	Thiagarajar College of Engineering, Madurai, India
Elena Lloret	Universidad de Alicante, Spain
Yoan Gutiérrez	Universidad de Alicante, Spain
Gandhi Hernandez	Universidad Tecnológica Metropolitana, Mexico
Manuel Sánchez-Rubio	Universidad Internacional de la Rioja, Spain
Mario Barcelo-Valenzuela	Universidad de Sonora, Mexico
Alonso Perez-Soltero	Universidad de Sonora, Mexico
Gerardo Sanchez-Schmitz	Universidad de Sonora, Mexico
Manuel Campos	Universidad de Murcia, Spain
Jose M. Juarez	Universidad de Murcia, Spain
José Luis Hernández Hernandez	Universidad Autónoma de Guerrero, Mexico
Mario Hernández Hernández	Universidad Autónoma de Guerrero, Mexico
Severino Feliciano Morales	Universidad Autónoma de Guerrero, Mexico
Guido Sciavicco	University of Ferrara, Italy
Ángel García Pedrero	Universidad Politécnica de Madrid, Spain
Miguel Vargas-Lombardo	Universidad Tecnologica de Panama, Panama
Denis Cedeño Moreno	Universidad Tecnologica de Panama, Panama
Viviana Yarel Rosales Morales	Instituto Tecnologico de Orizaba, Mexico
José Javier Samper-Zapater	Universidad de Valencia, Spain
Raquel Vasquez Ramirez	Instituto Tecnologico de Orizaba, Mexico
Janio Jadán Guerrero	Universidad Indoamérica, Ecuador

Local Organizing Committee

Katty Lagos-Ortiz	General coordinator, Universidad Agraria del Ecuador
Andrea Sinche Guzmán	Universidad Agraria del Ecuador
Cecilia Valle Lituma	Universidad Agraria del Ecuador
Tany Burgos Herrería	Universidad Agraria del Ecuador
Diego Arcos Jácome	Universidad Agraria del Ecuador
Jaime Cadena Iturralde	Universidad Agraria del Ecuador
Sirli Leython Chacón	Universidad Agraria del Ecuador
Winston Espinoza	Universidad Agraria del Ecuador
Ariadne Vega	Universidad Agraria del Ecuador
José Hernández	Universidad Agraria del Ecuador
Carlos Banchón	Universidad Agraria del Ecuador
Antonio Alava	Universidad Agraria del Ecuador

Fernando Damian	Universidad Agraria del Ecuador
Gloria Chave	Universidad Agraria del Ecuador
Karla Crespo	Universidad Agraria del Ecuador

Sponsoring Institutions

http://www.uagraria.edu.ec/

https://www.springer.com/series/11156

Contents

Contributors

Maritza Aguirre-Munizaga Escuela de Ingeniería en Computación e Informática, Facultad de Ciencias Agrarias, Universidad Agraria del Ecuador, Guayaquil, Ecuador
School of Computer Engineering, Faculty of Agricultural Sciences, Universidad Agraria del Ecuador, Guayaquil, Ecuador

Abel Alarcón-Salvatierra Facultad de Ciencias Matemáticas y Físicas, Universidad de Guayaquil, Cdla. Universitaria Salvador Allende, Guayaquil, Ecuador
Universidad Agraria del Ecuador, Guayaquil, Ecuador

Diego Arcos-Jácome Universidad Agraria del Ecuador, Guayaquil, Ecuador

Carlos Banchón Faculty of Agricultural Sciences, School of Environmental Engineering, Universidad Agraria del Ecuador, Guayaquil, Ecuador

Wilmer Baque-Bustamante Faculty of Agricultural Sciences, Computer Science Department, Agrarian University of Ecuador, Guayaquil, Ecuador

William Bazán-Vera Faculty of Agricultural Sciences, Agrarian University of Ecuador, Guayaquil, Ecuador
Universidad Agraria del Ecuador, Guayaquil, Ecuador

Oscar Bermeo-Almeida Faculty of Agricultural Sciences, Agrarian University of Ecuador, Guayaquil, Ecuador

Tamara Borodulina Faculty of Agricultural Sciences, School of Environmental Engineering, Universidad Agraria del Ecuador, Guayaquil, Ecuador

Tany Burgos-Herreria Universidad Agraria del Ecuador, Guayaquil, Ecuador

Roberto Cabezas-Cabezas Faculty of Agricultural Sciences, Agrarian University of Ecuador, Guayaquil, Ecuador

Jaime Cadena-Iturralde Agrarian Sciences Faculty, Agrarian University of Ecuador, Guayaquil, Ecuador

Kléber Calle Universidad Agraria del Ecuador, Guayaquil, Ecuador

Mario Cardenas-Rodriguez Faculty of Agricultural Sciences, Agrarian University of Ecuador, Guayaquil, Ecuador

Xavier Cárdenas-Rosales Agrosoft, Urdesa Central, Guayaquil, Ecuador

Cristhian Chávez Universidad Agraria del Ecuador, Guayaquil, Ecuador

Karla Crespo-León Agrarian Sciences Faculty, Agrarian University of Ecuador, Guayaquil, Ecuador

Elicia Cruz-Ibarra Faculty of Agricultural Sciences, Computer Science Department, Agrarian University of Ecuador, Guayaquil, Ecuador

Javier del Cioppo-Morstadt Faculty of Agricultural Sciences, Agrarian University of Ecuador, Guayaquil, Ecuador

Carlota Delgado-Vera Escuela de Ingeniería en Computación e Informática, Facultad de Ciencias Agrarias, Universidad Agraria del Ecuador, Guayaquil, Ecuador

Winston Espinoza-Moran Universidad Agraria del Ecuador, Guayaquil, Ecuador

Jesús Garcerán-Sáez Faculty of Computer Science, Department of Informatics and Systems, University of Murcia, Murcia, Spain

José Antonio García-Díaz Department of Information and Systems, Faculty of Computer Science, University of Murcia, Murcia, Spain

Francisco García-Sánchez Faculty of Computer Science, Department of Informatics and Systems, University of Murcia, Murcia, Spain

Mayra Garzón-Goya Escuela de Ingeniería en Computación e Informática, Facultad de Ciencias Agrarias, Universidad Agraria del Ecuador, Guayaquil, Ecuador

Freddy Gavilánez Universidad Agraria del Ecuador, Guayaquil, Ecuador

Raquel Gómez-Chabla Universidad Agraria del Ecuador, Guayaquil, Ecuador
Escuela de Ingeniería en Computación e Informática, Guayaquil, Ecuador

Erick González-Linch Facultad de Ciencias Matemáticas y Físicas, Cdla, Universidad de Guayaquil, Universitaria "Salvador Allende", Guayaquil, Ecuador

Paola Grijalva Escuela de Ingeniería en Computación e Informática, Guayaquil, Ecuador

Jorge Hidalgo Agrarian Sciences Faculty, Agrarian University of Ecuador, Guayaquil, Ecuador

Federico Murcia Labaña Faculty of Computer Science, Department of Informatics and Systems, University of Murcia, Murcia, Spain

Katty Lagos-Ortiz Escuela de Ingeniería en Computación e Informática, Facultad de Ciencias Agrarias, Universidad Agraria del Ecuador, Guayaquil, Ecuador
Facultad de Ciencias Matemáticas y Físicas, Cdla, Universidad de Guayaquil, Universitaria “Salvador Allende”, Guayaquil, Ecuador

Augusto Marcillo-Plaza Faculty of Agricultural Sciences, Computer Science Department, Agrarian University of Ecuador, Guayaquil, Ecuador

Tayron Martínez-Carriel Escuela de Ingeniería Agronómica, Facultad de Ciencias Agrarias, Universidad Agraria del Ecuador, Guayaquil, Ecuador
Faculty of Agricultural Sciences, Computer Science Department, Agrarian University of Ecuador, Guayaquil, Ecuador

Silvia Medina-Anchundia Facultad de Ciencias Matemáticas y Físicas, Cdla, Universidad de Guayaquil, Universitaria “Salvador Allende”, Guayaquil, Ecuador

José Medina-Moreira Facultad de Ciencias Agrarias, Universidad Agraria del Ecuador, Guayaquil, Ecuador
Facultad de Ciencias Matemáticas y Físicas, Cdla, Universidad de Guayaquil, Universitaria “Salvador Allende”, Guayaquil, Ecuador

Sergio Merchán-Benavides Escuela de Ingeniería Agronómica, Facultad de Ciencias Agrarias, Universidad Agraria del Ecuador, Guayaquil, Ecuador

Karen Mite-Baidal Faculty of Agricultural Sciences, Computer Science Department, Agrarian University of Ecuador, Guayaquil, Ecuador

César Morán Universidad Agraria del Ecuador, Guayaquil, Ecuador
Carrera de Ingeniería Agronómica, Facultad de Ciencias Agrarias, Universidad Agraria del Ecuador, Guayaquil, Ecuador

Manuel Flores Morán Facultad de Ciencias Matemáticas y Físicas, Universidad de Guayaquil, Cdla. Universitaria Salvador Allende, Guayaquil, Ecuador

José Ángel Noguera-Arnaldos Proyectos y soluciones tecnológicos avanzadas SLP (Proasistech), Murcia, Spain

Catherine Peralta Faculty of Agricultural Sciences, School of Environmental Engineering, Universidad Agraria del Ecuador, Guayaquil, Ecuador

Karina Real-Avilés Universidad Agraria del Ecuador, Guayaquil, Ecuador
Escuela de Ingeniería en Computación e Informática, Guayaquil, Ecuador

Tanya Recalde Carrera de Medicina, Facultad de Ciencias Médicas, Universidad de Guayaquil, Guayaquil, Ecuador

Isabel María Robles-Marín Proyectos y soluciones tecnológicos avanzadas SLP (Proasistech), Murcia, Spain

Alberto Ruiz Faculty of Computer Science, Department of Informatics and Systems, University of Murcia, Murcia, Spain

Andrea Sinche-Guzmam Escuela de Ingeniería en Computación e Informática, Facultad de Ciencias Agrarias, Universidad Agraria del Ecuador, Guayaquil, Ecuador

Evelyn Solís-Avilés Faculty of Agricultural Sciences, Computer Science Department, Agrarian University of Ecuador, Guayaquil, Ecuador

Mariuxi Tejada-Castro Escuela de Ingeniería en Computación e Informática, Facultad de Ciencias Agrarias, Universidad Agraria del Ecuador, Guayaquil, Ecuador

Rafael Valencia-García Department of Information and Systems, Faculty of Computer Science, University of Murcia, Murcia, Spain
Facultad de Informática, Universidad de Murcia, Campus Espinardo, Murcia, Spain

Mitchell Vásquez-Bermúdez Agrarian Sciences Faculty, Agrarian University of Ecuador, Guayaquil, Ecuador

Vanessa Vergara-Lozano Escuela de Ingeniería en Computación e Informática, Facultad de Ciencias Agrarias, Universidad Agraria del Ecuador, Guayaquil, Ecuador

Néstor Vera-Lucio School of Computer Engineering, Faculty of Agricultural Sciences, Universidad Agraria del Ecuador, Guayaquil, Ecuador

Intelligent and Knowledge-Based Systems

SePeRe: Semantically-Enhanced System for Pest Recognition

Jesús Garcerán-Sáez and Francisco García-Sánchez(✉)

Faculty of Computer Science, Department of Informatics and Systems, University of Murcia, 30100 Murcia, Spain
{jesus.garceran, frgarcia}@um.es

Abstract. Organic agriculture shows several benefits, namely, it reduces many of the environmental impacts of conventional agriculture, it can increase productivity in small farmers' fields, it reduces reliance on costly external inputs, and guarantees price premiums for organic products. However, its feasibility is often questioned due to the constraints in the use of chemical fertilizers and pesticides. For that reason, the general approach in organic agriculture is to deal with the causes of a problem (i.e., management practices aiming at preventing pests and diseases from affecting a crop) rather than treating the symptoms. In this work we propose a fully-fledged, integral, comprehensive technological solution for the early detection of plant diseases and pests, and the suggestion of organic agriculture-compliant treatments. A proof-of-concept prototype has been developed that identifies the presence of harmful conditions in the crop and lists appropriate treatments.

Keywords: Organic agriculture · Pest recognition · Ontology

1 Introduction

With the rapid increase in the world's population [1], there is an ever-growing need for a sustainable food supply. According to experts, fruits and vegetables constitutes the basis for a healthy diet [2]. However, in the last few years the worldwide dispersion of virulent plant pests and diseases has caused significant decreases in yield and quality of crops, in particular fruit, cereal and vegetables [3]. Climate change and the intensification of global trade flows further accentuates the issue. Integrated Pest Management (IPM) is an approach to pest control that aims at maintaining pest insects at tolerable levels, keeping pest populations below an economic injury level [4]. According to the Health and Food Safety department of the European Commission, IPM '*means careful consideration of all available plant protection methods and subsequent integration of appropriate measures that discourage the development of populations of harmful organisms and keep the use of plant protection products and other forms of intervention to levels that are economically and ecologically justified and reduce or minimize risks to human health and the environment*' [5].

Organic agriculture is one of the best practices in ensuring environmental sustainability, soils fertility and biodiversity while providing a secure source of healthy food for the growing population [6]. Moreover, it has a strong potential for building

R. Valencia-García et al. (Eds.): CITAMA 2019, AISC 901, pp. 3–11, 2019.
https://doi.org/10.1007/978-3-030-10728-4_1

resilience in the face of climate change. However, its feasibility is often questioned due to the constraints in the use of chemical fertilizers and pesticides. Consequently, IPM strategies should be revised and adapted to the limitations imposed by organic agriculture regulations [7]. In particular, the early identification of pests and diseases becomes mandatory to allow the adoption of preventive measures. Yet, farmers have very limited resources to detect the outbreaks trigger factors and act accordingly. Moreover, organic agriculture-compliant treatments are still unknown for most people. It is thus necessary to provide farmers with the means to first recognize the presence of diseases and pests in the plant, and second develop preventive actions and use IPM practices allowed for organic production, in order to limit their harmful effects.

In this work, we present the first step towards a fully-fledged, semantically-enhanced decision support system for integrated pest management in organic agriculture. While the ultimate goal is to build a complete agricultural knowledge base by gathering data from multiple, heterogeneous sources and to develop tools to assist farmers in decision making concerning pests and diseases control, in this work we describe an initial prototype of the system that relies on a relational database and takes advantage of Google's Cloud Vision API[1] to identify plagues of crop-destroying pests from the photos taken by the farmer. Currently, the application focuses on the crops with the most economic value in the Mediterranean countries, more specifically the southeastern regions of Spain, such as olives, grapes, citrus fruits (e.g., oranges and lemons), tree-nuts (e.g., almonds), and vegetables (e.g., tomatoes,) [8]. However, it has been built so as to be easily extensible to further crops. Accessibility and user-friendliness have been emphasized in the development of the system since it is geared toward non-technical users.

The rest of the paper is organized as follows. In Sect. 2, other tools aimed at assisting in the recognition of pests' outbreaks are analyzed and the benefits a semantic approach highlighted. The framework proposed in this work is described in detail in Sect. 3 and the prototypical implementation of the first step towards that outcome is presented in Sect. 4. Finally, conclusions and future work are put forward in Sect. 5.

2 Related Work

Nowadays, the use of new technologies in agriculture is almost pervasive across the world [9]. Smart farming refers to using high-tech farming techniques and technologies to improve production output while minimizing cost and preserving resources [10]. The main applications of ICT in agriculture (also known as 'e-agriculture'[2]) include the use of GPS and Geographic Information Systems (GIS) for precision farming, smartphone apps for e-learning and crop management, RFID for product tracking, knowledge management systems with information and best practices, etc. New technologies promote sustainable agricultural development and food security by improving the use of information, communication, and associated technologies in the sector. The benefits

[1] https://cloud.google.com/vision/.
[2] http://www.fao.org/e-agriculture/.

of the application of ICT in agriculture are (i) timely and updated information on agriculture related issues, (ii) increased efficiency, productivity and sustainability, (iii) better marketing exposure, (iv) inputs optimization and, thus, risk reduction, and (v) improved networking and communication, among others. In this section, both pest recognition tools and the application of semantic technologies in agriculture are analyzed in some detail.

Concerning pest recognition, a number of different image processing tools can be found in literature [11–13]. In [11] the authors propose a content-based image retrieval system that given an image, returns the images in the database most alike. It makes use of color, shape and texture features of leaf reaching an average precision of 80% in their tests with diseased soybean leaves. The authors in [12] claim that assigning the same weight to all features in the classification process of existing image-based crop disease recognition results in reduced precision. They propose the use of a sparse representation technique that helps in reducing computation cost while improving the recognition performance. Their tests in cucumber disease recognition provides an overall rate of 85.7%. A genetic algorithm is used in [13] to perform image segmentation and classify the leaf diseases. Then color and textual features of leaves are employed to recognize the diseases by leveraging soft computing techniques such as the Minimum Distance Criteria and Support Vector Machines. They tested their approach on different plants with disparate diseases reaching an overall accuracy of 95.71%. Most of the existing applications require the use of very sophisticated photography equipment not accessible to smallholders. Besides, the precision of the approaches strongly depends on the quality and completeness of the database. The tool described in this manuscript takes advantage of the collective intelligence by using Google's Cloud Vision API, which consider all Internet-reachable image resources.

Semantic technologies are also being applied in the agriculture sector [14, 15]. Ontologies are a useful means to integrate and harmonize data from different sources, and facilitate inferring and reasoning over the shared conceptualization. AgroPortal[3] [14] is an ontology repository for the agronomy domain that allows for identifying, hosting and using the many vocabularies and ontologies in this field. Over 100 ontologies are currently hosted in this repository with more than 1.7 million classes and 1.9 million individuals. Agricultural technology, animal science, ecology, farming systems, etc., are all covered in this reference repository. In [15] the authors describe a knowledge-based system to support the diagnosis of plant diseases. The system rests on a rule-based engine built with the assistance of domain experts. If the symptoms described by the farmer trigger a rule, then a diagnose is provided, and relevant treatments and recommendations are suggested to the farmer. It is not clear the way symptoms are inputted into the system, but it relies on the perception of the farmer. In this work we propose an application that makes use of pictures to automatize symptoms gathering.

---

[3] http://agroportal.lirmm.fr/.

3 Proposed Framework

The main goal of this ongoing work is to empower end users with the knowledge necessary to develop preventive actions and to use integrated pest management practices that are allowed for organic production. To this end, relevant data coming from different sources is integrated into a knowledge base. Then, different expert system tools are enabled for farmers to exploit that knowledge. The components that form part of the overall environment are depicted in Fig. 1. In a nutshell, the system works as follows. Different ontology population techniques are used to gather information from heterogeneous (either structured, semi-structured or unstructured) data sources. Knowledge entities complying with an ontology schema (i.e., the Agricultural Ontology) are thus created in the knowledge base. Different tools rely on the knowledge base to (i) analyse its content to understand and foresee outbreaks, and (ii) identify diseases and pests in the plant and suggest the most adequate treatments. To support this, predictive analysis techniques along with image processing tools and natural language processing methods are put into place. Next, some details about the constitutive parts of the framework are provided.

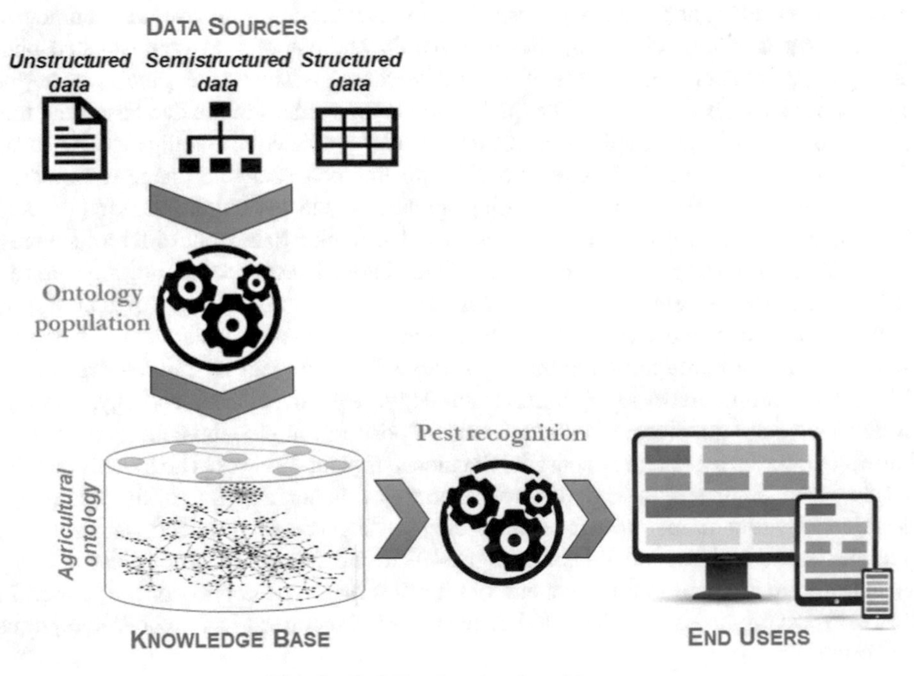

Fig. 1. SePeRe functional architecture

3.1 Agricultural Ontology and Ontology Population

Many factors (from meteorological conditions to physiological factors) could contribute to the outbreak of plant diseases and pests, and analyzing the large amounts of data related to all possible involved conditions is a very challenging endeavor. The use of semantic models partially alleviates the big data analytics burden by dealing with information at the knowledge level. The backbone of this proposal rests on a semantic conceptualization of the plant pests and diseases domain, which allows (i) the representation of the main concepts involved and their relationships, (ii) the seamlessly integration of data from heterogeneous sources, and (iii) the inferencing of new implicit knowledge by means of reasoning processes.

The ontology model that will sustain the knowledge base will be constructed from scratch by following the Linked Open Data best practices. In the development of this knowledge base different inputs will be considered, namely, (i) the knowledge of the domain expert partners, (ii) existing ontologies in the agronomic domain, and (iii) official data sources linked to the plant health domain. Other commonly used ontologies and vocabularies, such as those publicly available in the AgroPortal repository, will be examined so as to make use of internationally accepted terms and enable interoperability, data integration and standardization. The most up-to-date information about plants, diseases, pests, organic production, etc., published in official sites such as the European Food Safety Authority or the Organic Agriculture Programme of the Food and Agriculture Organization will be used to complete our knowledge model. In particular, in order for the knowledge base to be useful in problem identification, the characterization factors such as problem-causing pathogens, affected area (hectares), affected crops, yield losses, or infection timing, should be included. Other common concepts that are relevant in this domain and will be included in the semantic model are plant variety, irrigation water flow, distance between plants, temperature, wind direction, insect populations, irrigation dates, treatment.

Historical data publicly available worldwide in open (linked) data sources will be exploited to populate the knowledge base (e.g., from general purpose resources such as DBpedia[4] or the EU Open Data Portal[5] to meteorological data such as AEMET[6] or more specific agricultural thesaurus such as AGROVOC[7]). For this, different mechanisms for ontology population from unstructured, semi-structured and structured data sources will be explored. In this context, it is also important to take into account that plant diseases are abnormal dysfunctions that can evolve in such a way that every year new types are coming up. Hence, as part of this component we will also define strategies capable of updating the structure, concepts and properties in the knowledge model. Ontology evolution techniques will be used to keep the knowledge base continuously up-to-date.

---

[4] http://wiki.dbpedia.org/.

[5] http://data.europa.eu/euodp/.

[6] http://www.aemet.es/en/datos_abiertos.

[7] http://aims.fao.org/standards/agrovoc.

3.2 Pest Recognition

On top of the knowledge base, three expert systems assist in (i) the definition of optimal preventive actions, and (ii) the early identification of pests and diseases, covering a widespread spectrum of evidentiary inputs, namely, images, texts and environmental parameters. Problem-causing pathogens, affected area (hectares), affected crops, yield losses or infection timing are some of the factors that can help characterize the problem source. Machine learning algorithms will be used to thoroughly explore the relationships between the variables in the model and find out the conditions under which certain crop diseases and pests break out. The analysis of real time data from crop monitored parameters can then be used to recommend necessary preventive or remedial treatments. On the other hand, an integrated approach for the identification of the cause of damage conditions in plants will be put into place comprising both natural language processing methods and image processing techniques. The natural language textual description of the situation and photos of the externally observable symptoms constitute the inputs of these sub-systems, which should have been previously trained. The outcome of this component is a list of likely plant diseases and pests along with their relative probability of occurrence and the corrective measures as suggested by organic agriculture regulations and integrated pest management practices. In order to correctly identify the pests/diseases affecting a plant, the scores returned by all three sub-systems will be aggregated. The precision of these components will improve over time with feedback from expert stakeholders.

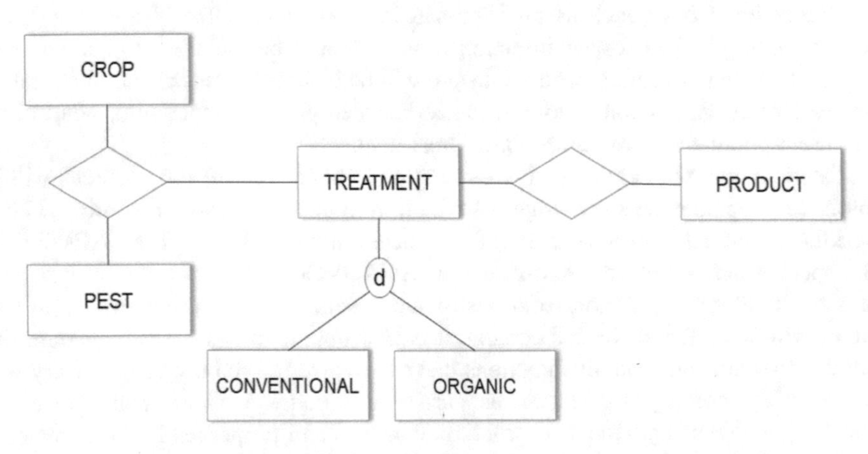

Fig. 2. Database schema

4 Proof-of-Concept Implementation

With the aim of having an initial proof-of-concept implementation of the proposed framework and a well-designed skeleton for future developments, we built a responsive web application capable of identifying plant diseases from images and suggesting the most appropriate treatment. Instead of employing an ontology-based knowledge management system, this prototype rests on a relational database which contains information about the most important diseases and pests that affect common Mediterranean crops, such as olives, almonds, grapes, etc. The conceptual schema of the database is depicted in Fig. 2. The main entities considered are *'Crop'*, which refers to the cultivated produce of the ground; *'Pest'*, which refers to a plant or animal detrimental to crops, including plant diseases; *'Treatment'*, which refers to both conventional and organic pest control strategies, including cultural, chemical and biological means; and *'Product'*, which refers to the commercial products containing the active substances defined in *'Treatment'*.

The application, which is accessible from any device, receives the name of the crop and a picture (e.g., a photo taken by the farmer) as input (see Fig. 3 [left]), and provides information about the condition affecting the plant and the most appropriate treatments as output (see Fig. 3 [right]). For this, Google's Cloud Vision API is leveraged. The received picture is annotated using this API and the resulting annotations are matched against the information in the local database. If a tag if found associated to the picture that matches the name of a pest or disease in the database, then the system concludes that the referred pest is present, and the user is shown information about it, its (both conventional and organic) treatments and existing products to deal with the pest.

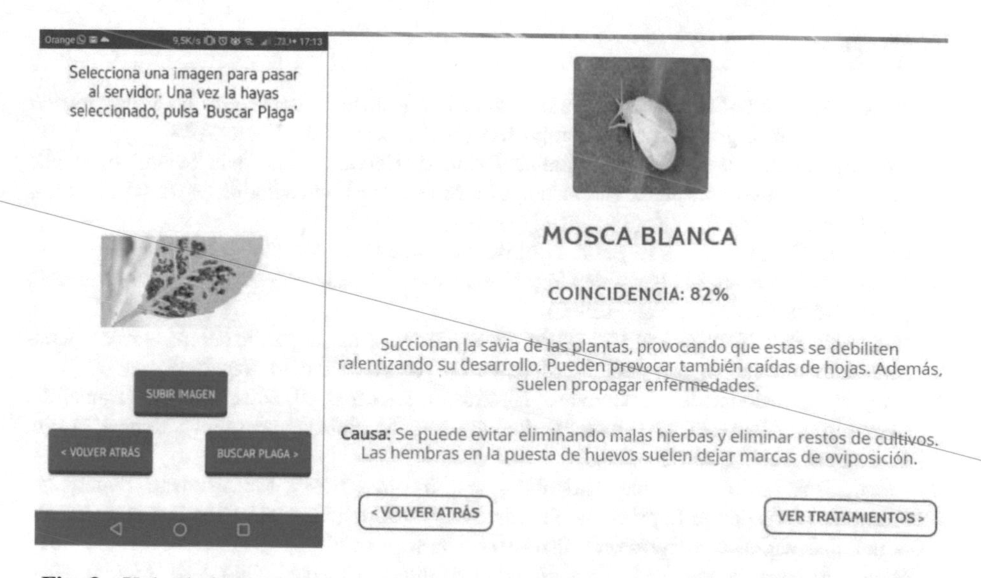

Fig. 3. Upload photo (Android app) [left]; Pest identification (Web app) [right]; in Spanish

5 Conclusions and Future Work

Organic farming is based on the optimal use of natural resources and the avoidance of most synthetic inputs (e.g., fertilizers, pesticides, etc.), thus producing organic food while conserving the fertility of the land and respecting the environment. The overall objective of this project is the development of a semantically-empowered system for the early detection of plant diseases and pests and the recommendation of preventive measures complying with organic agriculture regulations. Embedded experts' knowledge along with machine learning methods constitute the core of the system, which is fed with evidence data from different, heterogeneous sources and produces a set of preventive practices, as well as control practices using biological, mechanical control and natural pesticides as required in organic agriculture.

A first step has been taken towards this grand challenge; a prototypical proof-of-concept tool has been developed that, given a picture, identifies the presence of certain plant pests and diseases, suggesting appropriate treatments and products. We plan to extend this initial prototype by adding the components envisioned in the proposed framework. In particular, semantic technologies will be leveraged to facilitate the integration of heterogeneous data originating from disparate sources, and its analysis. The formal underpinnings of ontologies enables tasks like logical reasoning and instance classification that can yield additional benefits for semantic integration.

Acknowledgements. This work has been supported by the Spanish National Research Agency (AEI) and the European Regional Development Fund (FEDER/ERDF) through project KBS4FIA (TIN2016-76323-R).

References

1. Roser, M., Ortiz-Espina, E.: World population growth. Our World in Data. https://ourworldindata.org/world-population-growth (2017). Accessed 14 Aug 2018
2. Healthy Eating Plate & Healthy Eating Pyramid: Harvard T.H. Chan School of Public Health. https://www.hsph.harvard.edu/nutritionsource/healthy-eating-plate/ (2018). Accessed 14 Aug 2018
3. Oerke, E.-C.: Crop losses to pests. J. Agric. Sci. **144**(1), 31–43 (2006)
4. Prokopy, R., Kogan, M.: Integrated Pest Management. In: Encyclopedia of Insects, 2nd edn., pp. 523–528. Academic Press (2009)
5. Integrated Pest Management (IPM): European Commission. https://ec.europa.eu/food/plant/pesticides/sustainable_use_pesticides/ipm_en (2018). Accessed 14 Aug 2018
6. Badgley, C., Moghtader, J., Quintero, E., Zakem, E., Chappell, M.J., Avilés-Vázquez, K., Samulon, A., Perfecto, I.: Organic agriculture and the global food supply. Renew. Agric. Food Syst. **22**(02), 86–108 (2007)
7. Baker, B.P., et al.: Organic Agriculture and Integrated Pest Management: Synergistic Partnership Needed to Improve the Sustainability of Agriculture and Food Systems. https://organicipmwg.files.wordpress.com/2015/07/white-paper.pdf (2018). Accessed 14 Aug 2018
8. García Álvarez-Coque, J.M.: La agricultura mediterránea en el siglo XXI. Mediterráneo Económico. Colección estudios socioeconómicos, pp. 1–312. Instituto de Estudios de Cajamar, Almería, Spain (2002)
9. Woodard, J., et al.: ICT in Agriculture (Updated Edition): Connecting Smallholders to Knowledge, Networks, and Institutions. The World Bank (2017)
10. Boschetti, M., Schoitsch, E.: Smart Farming - Introduction to the Special Theme. ERCIM News **2018**(113) (2018)
11. Patil, J.K., Kumar, R.: Analysis of content based image retrieval for plant leaf diseases using color, shape and texture features. Eng. Agric. Environ. Food **10**(2), 69–78 (2017)

12. Zhang, S., Wu, X., You, Z., Zhang, L.: Leaf image based cucumber disease recognition using sparse representation classification. Comput. Electron. Agric. **134**, 135–141 (2017)
13. Singh, V., Misra, A.K.: Detection of plant leaf diseases using image segmentation and soft computing techniques. Inf. Process. Agric. **4**(1), 41–49 (2017)
14. Jonquet, C., et al.: AgroPortal: a vocabulary and ontology repository for agronomy. Comput. Electron. Agric. **144**, 126–143 (2018)
15. Lagos-Ortiz, K., Medina-Moreira, J., Paredes-Valverde, M.A., Espinoza-Morán, W., Valencia-García, R.: An ontology-based decision support system for the diagnosis of plant diseases. J. Inf. Technol. Res. **10**(4), 42–55 (2017)

Selecting Safe Walking Routes to Minimize Exposure Time in Outdoor Environments

José Antonio García-Díaz[1](✉), José Ángel Noguera-Arnaldos[2], Isabel María Robles-Marín[2], Francisco García-Sánchez[1], and Rafael Valencia-García[1]

[1] Department of Information and Systems, Faculty of Computer Science, University of Murcia, 30100 Murcia, Spain
{jose-antonio.garcia8, frgarcia, valencia}@um.es

[2] Proyectos y soluciones tecnológicos avanzadas SLP (Proasistech), Avda. Primero de Mayo, nº2, Torres Azules. Torre A, 4ª Planta Drcha, Murcia, Spain
{jnoguera, isabel.robles}@proasistech.com

Abstract. Walking is a beneficial activity both for the environment and for one's health. However, the safety of pedestrians may be compromised due to the presence of harmful substances in the atmosphere, areas with poor light conditions, or physical barriers among other factors. In this work we propose a solution that takes into account the risk factors affecting users with the multiple hazards detected in outdoor environments. Hazards are identified by gathering data from heterogeneous sources, such as a network of air-quality monitoring stations and open-data sources. The developed software component has been attached to the *AllergyLESS* platform, a recommender system that provides safe routes recommendations. In addition, a field test was carried out to test the effectiveness of the system in a real environment with successful results.

Keywords: Information systems · Decision support · Multi-objective shortest path · Routing problem

1 Introduction

According to the World Health Organization, a pedestrian is any person who makes, at least, one part of their journey by walking [1]. Walking is a beneficial activity both for the environment and for health [2]. However, during the journey, the safety of pedestrians may be compromised due to: (i) environmental factors, such as air pollution [3] and natural disasters [4]; (ii) traffic issues, such as traffic jams [5] and traffic accidents [1]; and (iii) sudden phenomena, such as urban floods [6]. Therefore, there is an increasing interest in the development of software applications that take advantage of real-time information to promote the safety of pedestrians.

This paper describes the development of a software module responsible of selecting safe routes in outdoor environments. The proposed solution takes under consideration (i) the different nature of hazards, such as the presence of allergens in the atmosphere, poor light conditions or physical barriers; (ii) the distance between users and the

R. Valencia-García et al. (Eds.): CITAMA 2019, AISC 901, pp. 12–19, 2019.
https://doi.org/10.1007/978-3-030-10728-4_2

existing hazards in each alternative route; and (iii) a list of the hazards that affects each user the most.

The findings of this study were added to the *AllergyLess* platform [7], a routing recommender system that provides safe routes by minimizing the exposure time to allergens in smart cities. The main goal is to use real-time information concerning hazards in outdoor environments to provide recommendations for harmless alternative routes that help pedestrians to safely reach their destination.

The remainder of this paper is organized as follows. In Sect. 2, some similar applications regarding pedestrian safety are analysed. In Sect. 3, the system architecture is described, emphasizing the strategy to locate the air-quality monitoring stations and the way in which the proposed solution handles multi-criteria environments. The experiments performed to validate this module are shown in Sect. 4. Finally, conclusions and some further lines of research are put forward in Sect. 5.

2 Related Work

In the bibliography it is possible to find several tools related to the safety of pedestrians. However, the majority of applications found are focused only in preventing traffic accidents. *WalkSafe* [8] and *CarSafe* [9], for example, are apps that assists users in avoiding traffic accidents by making use of the back camera of their mobile phones while walking on the streets. *LookUp* [10] relies in the use of a sensor installed in the shoes of the user to detect step patterns, inclination and surface characteristics. The gathered information is transferred to the smartphone of the user and the information is analyzed to determine patterns that involve risk situations for the pedestrians. Other solutions explore the communication between cars and pedestrians to promote road safety of the pedestrians. In [11] the authors stablished the requirements for the minimum information exchange and distance needed to transmit information in order to be effective to avoid traffic accidents.

Concerning air-pollution and allergens, a similar approach of this work was conducted in the city of Valencia, Spain with the app R-ALERGO [12] which "uses modified grid calculations where impedances have been calculated based on the variables that affect allergens exposure: temperature, humidity and wind direction, pollution level" among other metrics.

3 System Architecture

The *AllergyLESS* platform [7] is a mapping mobile application that shows high-concentrations of allergens and other substances related to air quality. The presence of allergens is detected by a network of air-quality monitoring stations. It also calculates which is the shortest route in a trip while maximizing the distance to the detected threats. The system architecture is organized following a three-tier architecture as shown in Fig. 1.

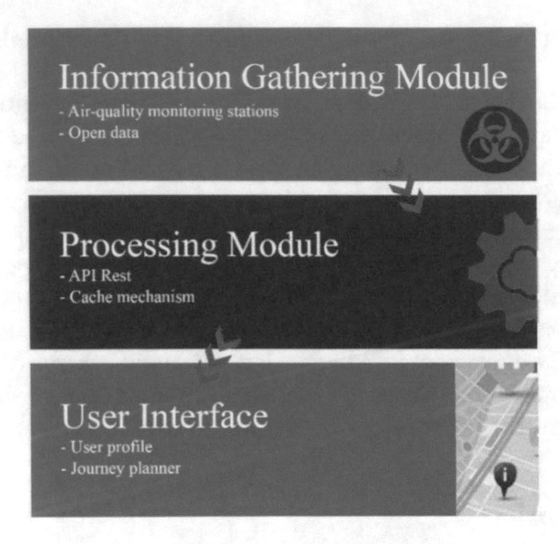

Fig. 1. System architecture

First, the information gathering module is responsible for the acquisition, filtering and normalization of data collected from a network of air-quality monitoring stations. This network detects the presence of pollen and other substances concerning air quality, such as ozone, carbon monoxide and sulphur dioxide among others. These substances were extracted due to their relation to allergens and asthma [13, 14]. A detailed list of these parameters is shown in Table 1. In addition, meteorological data are gathered from open data sources, such as *AEMET*[1] or *OpenWeatherMap*[2], and used to forecast the presence of allergens in scenarios where the number of available air-quality monitoring stations is limited. Second, the processing module is in charge of providing an access/entry-point to the front-end application. This module was designed following a restful architecture [15]. This architecture, which relies on the HTTP protocol, allows the easy implementation of cache mechanisms to reduce the queries overload. Finally, the user interface consists in a web application where users with can view a map with detailed information about the detected hazards. In this module, the users can use the journey planner to obtain information regarding safe alternative routes for their journey.

To method by which the system is capable of analyzing the safety of each route from the hazards in the environment is presented in the following subsections. First, the strategy to locate the air-quality monitoring stations to cover the largest area possible is described. Second, the parameters that model each possible hazard are identified. Finally, a benchmark of different third-party journey planner systems is put forward and we point out what the needed requirements are.

---

1 http://www.aemet.es/es/portada.

2 https://openweathermap.org/.

Table 1. Air quality parameters measured by sensors

Sensor type	Symbol	Ranges
Ozone	O_3	0–5 ppm
Carbon monoxide	CO	20–2000 ppm
Nitrogen dioxide	NO_2	0–5 ppm
Carbon dioxide	CO_2	10–1000 ppm
Sulfur dioxide	SO_2	20–2000 ppm
Fine particles	$PM_{2.5}$	0–1.5 V/(0.1mg/m^3)
Large particles	PM_{10}	0–35 pcs/cm^3
Humidity	Hum	0–100 %
Temperature	°C	0–100 °C

3.1 Location of the Air-Quality Monitoring Stations

The air-quality monitoring stations are PC-embedded nodes with sensors attached to them. Each sensor is equipped with a receptor that analyses the atmosphere to detect the presence of certain air quality-related substances. Considering that each air-quality monitoring station has a fixed coverage ratio, the strategy to locate them in an outdoor environment was studied from a sensor network perspective [16].

As the number of available air-quality monitoring stations may be limited, it is necessary to place the sensors covering the maximum possible area. This problem is known as the *Set Cover Problem* and, formally, consists in the selection of the minimum number of subsets whose union has maximum cardinality [17]. The technique applied in the *AllergyLESS* prototype consisted in applying Voronoi diagrams to find the combination of subsets with the largest coverage area. This technique has been previously suggested in a pedestrian flow-related research [18].

3.2 Nature of the Barriers

Apart from the hazardous substances present in the atmosphere, the system also takes into consideration other types of threats, such as poorly illuminated areas, narrow pavements, high curbs, etc. For each hazard, the platform keeps a record of (i) its geographical coordinates, (ii) a time-interval, (iii) an action radius, and (iv) its intensity level. Associating a time-interval parameter to the hazards is useful to indicate what time the hazard is active; this is beneficial to, for example, mark areas with poor lighting conditions when relevant. On the other hand, attaching an action radius to the hazards is useful to set a fuzziness level. For example, when tracking pollen, it is very likely to find similar pollen levels in the surrounding area [19]. In contrast, when dealing with physical barriers, such as a road under construction that has been temporally closed, its action radius will be a discrete value. Finally, intensity can be defined as the level in which the hazard affects each user and it is closely related to his/her profile (i.e., each user should establish an intensity threshold for each given hazard). An unsolvable barrier would have an intensity level close to 1.

Once gathered all the information about the surrounding hazards, their time intervals and radius, the system can calculate the routes minimizing the overall issues. For this, it is also necessary to have enough knowledge about users and the environmental factors most affecting them, such as pollen [20] or dust [21] among others. When users first enter the application, they are asked to complete their profile. This profile information can be updated later at any time. Once the profile has been entered and access to the device's geolocation has been granted, a map containing the local area surrounding the user is shown. Then, users are asked to introduce the address or place they want to get to. Once the destination has been set, the platform tracks the movements of the user by using the browser's Geolocation API. In order for the system to be aware of the changes in the environment in real time, it continuously calculates the best alternative route from the users' current position and their final destination.

Selecting safe routes in hazardous environments is a multicriteria decision problem similar to the shortest path problem [22], which consists in finding the path between two nodes in a graph minimizing the sum of the weights of each node. In most cases, problems of these kinds need to reach a compromise solution that manages to minimize the risk even if they are not optimal solutions for each subproblem [23]. However, the main difficulties associated to the problem at hand include the following: (i) the weights of each node can be influenced by each user's personal conditions, and (ii) the presence of the hazards is not always clear due to difficulties while detecting its occurrence and should be dealt with as a fuzzy problem. To solve these issues, the platform labels each route with a score based on their proximity to the detected hazards. The lower the score, the safer the route is. As commented above, every type of hazard has an associated radius of action, so hazards that do not collapse with the routes are dismissed.

The system is prepared to handle multiple urban areas with the same database. To limit the number of calculations, the system only loads the hazards in the same area of the user. Later, when the system knows what the path of the user and the alternative routes is, the hazards that are out of a safe distance of them are removed. This safe distance was set arbitrarily based on the experiments and the validation of system as 1 kilometre. That is because no reference value has been found in the literature except in [24], which indicates a maximum distance of 100 kilometres.

An incremental approach has been followed to calculate the safety of the journey. First, the distance is calculated as a straight line between the origin and the destination, and so the complexity lies in the number of hazards detected by the monitoring stations within the radius of the secure distance. The distance of each hazard to the straight line is calculated by applying the Haversine Formula [25]. Second, each alternative route is analysed individually. For this, each route is divided into segments and their distance to each hazard is calculated individually. The distance is weighted by multiplying it with a factor between 0 and 1 that represents the impact of a particular hazard on the user according to his/her profile. For example, pollen and dust sensitive users will have a weight of 1 for these hazards.

3.3 Journey Planner

The detection of alternative routes was performed by using a third-party journey planner. This type of applications is used to find optimal means to travel between two or more given locations. *AllergyLESS* platform required a journey planner that supports routes on foot and also returns more than one alternative route to reach to the destination. Additionally, it would be useful for the journey planner to provide elevation information of the routes. The benchmark of the alternatives concerning the third-party journey planner is shown in Table 2. The studied platforms were: (i) *Google Maps,* due to its simplicity and high availability; (ii) Yours[3], a route system for *OpenStreetMap* [26]; and (iii) *OpenRouteService* [27], which provides a direction service including gradients, suitability, different surface types among other characteristics. Out of the three, the selected journey planner was Google Maps because it was the only choice that provided real support to alternative routes.

Table 2. Benchmark of different routing system

Routing System	Free usage	Global coverage	Alternative routes	Walking routes	Elevation information
Google Maps	Yes	Yes	Yes	Yes	External
Yours	Yes	Yes	Two	Yes	No
Open Route Service	Yes	Yes	Planned	Yes	Yes

4 Experiment

Two different testbeds comprise this experiment, namely, a simulated environment experiment and a field experiment. In the simulated experiment, the database was randomly populated with pollen measures, physical barriers and poor streetlight zones within an urban area. In each test of this experiment, random values are assigned to a dummy user's profile. Then, the system requests to the journey planner all the alternative routes between two fixed points within the urban area considered and assigns to each of them a score based on the user's profile values for the sensitivity to pollen and the other random hazards in the area nearby. Once the all the routes are calculated and a score is assigned to each of them, the score was analysed to check if it was correct. Also, the suggested best routes were individually verified to check whether the main hazards were avoided or not. This experiment allowed to adjust the weighting factor associated to each element in the user profile to find a balance between the duration of the journey and the distance to each hazard. Once this experiment was concluded, a field test was performed. Some university volunteers were asked to perform a daily routine including a walk across the Campus for a distance of about 1.3 km each day for a period of 5 days. This experiment was designed in a similar fashion similar to the original *AllergyLESS* platform [7] testbed but in this case including other hazards with different intensity values and effect radius.

[3] http://yournavigation.org/.

5 Conclusions and Future Work

In this paper, a method for safe path selection in a multi hazardous environment is described. The proposed approach has been integrated into the *AllergyLESS* platform and tested in order to select the safest routes when multiple events can take place at the same time. The results showed that it is possible to find routes that provide the proper compromise between the health status of the user and time constraints. In addition, each analysed route is assigned a 'safe' score; if no alternative route reaches the threshold score, the system warns the user that it is better to consider alternative transportation methods, such as public transport.

At the time of this experiment, *Open Route Service* is not returning more than one route. As its API is prepared to do it and this journey planner offers more complete information such as elevation points, we are analysing the possibility of switch from *Google Maps* to *OpenRoute Service* in the future. The evaluation information would be useful to calculate the elapsed time of each route more accurately and it is an important factor to consider when dealing with people with respiratory issues.

Acknowledgments. This work has been supported by the Comunidad Autónoma de la Región de Murcia (CARM) and the European Regional Development Fund (FEDER/ERDF) through project 2I16SAE00025 under the RIS3MUR program.

References

1. WHO: Helmets: a road safety manual for decision-makers and practitioners (2006)
2. Lee, I. M. Buchner, D. M.: The importance of walking to public health. Medicine and science in sports and exercise, (7 Suppl), S512-8 pp. 40 (2008)
3. Künzli, N., Kaiser, R., Medina, S., Studnicka, M., Chanel, O., Filliger, P., Schneider, J.: Public-health impact of outdoor and traffic-related air pollution: a European assessment. The Lancet **356**(9232), 795–801 (2000)
4. Wood, N., Jones, J., Schmidtlein, M., Schelling, J., Frazier, T.: Pedestrian flow-path modeling to support tsunami evacuation and disaster relief planning in the US Pacific Northwest. Int. J. Disaster Risk Reduct. **18**, 41–55 (2016)
5. Shi, Q., Abdel-Aty, M.: Big data applications in real-time traffic operation and safety monitoring and improvement on urban expressways. Emerging Technologies **58**, 380–394 (2015)
6. Russo, B., Gómez, M., Macchione, F.: Pedestrian hazard criteria for flooded urban areas. Natural hazards **69**(1), 251–265 (2013)
7. García-Díaz, J.A., Noguera-Arnaldos, J.A., Hernández-Alcaraz, M.L., Robles-Marín, I.M., García-Sánchez, F., Valencia-García, R.: AllergyLESS. An intelligent recommender system to reduce exposition time to allergens in smart-cities. In: DCAI'18 Distributed Computing and Artificial Intelligence, pp. 61–68. Toledo, Spain (2018)
8. Wang, T., Cardone, G., Corradi, A., Torresani, L., Campbell, A.T.: WalkSafe: a pedestrian safety app for mobile phone users who walk and talk while crossing roads. In: HotMobile'12, Workshop on Mobile Computing Systems & Applications, pp. 5 (2012)
9. You, C.W., et al.: Carsafe app: alerting drowsy and distracted drivers using dual cameras on smartphones. In: AMC Proceeding of the 11th Annual International Conference on Mobile Systems, Applications, and Services, pp. 13–26 (2013)

10. Jain, S., Borgiattino, C., Ren, Y., Gruteser, M., Chen, Y., Chiasserini, C.F.: Lookup: enabling pedestrian safety services via shoe sensing. In: AMC Proceedings of the 13th Annual International Conference on Mobile Systems, Applications, and Services, pp. 257–271 (2015)
11. Anaya, J.J., Merdrignac, P., Shagdar, O., Nashashibi, F., Naranjo, J.E.: Vehicle to pedestrian communications for protection of vulnerable road users. In: IEEE Intelligent Vehicles Symposium Proceedings, pp. 1037–1042 (2014)
12. Temes Cordovez, R.R., Hernández Fernández de Rojas, D., Moya Fuero, A., Martí Garrido, J.: APP R-ALERGO: allergy-healthy routes in Valencia. In: Back to the Sense of the City: International Monograph Book. pp. 1095–1105 (2016)
13. Oprea, M. M: AIR_POLLUTION_Onto: an ontology for air pollution analysis and control. In: IFIP International Conference on Artificial Intelligence Applications and Innovations, pp. 135–143. Boston (2009)
14. IARC Working Group on the Evaluation of Carcinogenic Risks to Humans: Outdoor air pollution measurement methods. IARC Monographs on the Evaluation of Carcinogenic Risks to Humans **109**, p. 9 (2016)
15. Richardson, L. Ruby, S.: RESTful web services, O'Reilly Media, Inc. (2008)
16. Wang: Coverage problems in sensor networks: a survey. ACM computing surveys, **43**(4), pp. 32 (2011)
17. Feige, U.: A threshold of ln n for approximating set cover. J. ACM **45**(4), 634–652 (1998)
18. Xiao, Y., Gao, Z., Qu, Y.: Li, X: A pedestrian flow model considering the impact of local density: Voronoi diagram-based heuristics approach. Trans. Res. Part C: Emerg. Technol. **68**, 566–580 (2016)
19. Austerlitz, F., Dick, C.W., Dutech, C., Klein, E.K., Oddou-Muratorio, S., Smouse, P.E., Sork, V.L.: Using genetic markers to estimate the pollen dispersal curve. Mol. Ecol. **13**(4), 937–954 (2004)
20. Bennett, K.D., Willis, K.J.: Pollen. Tracking environmental change using lake sediments. In: Developments in Paleoenvironmental Research pp. 5–32 (2002)
21. Boulet, L.P., et al.: Comparative degree and type of sensitization to common indoor and outdoor allergens in subjects with allergic rhinitis and/or asthma. Clin. Exp. Allerg. **27**(1), pp. 52–59 (1997)
22. Osyczka, A.: Multicriteria optimization for engineering design. Design Optimization, pp. 193–227 (1985)
23. Ehrgott, M.: Multicriteria optimization (491). Springer Science & Business Media, Berlin (2005)
24. Spangl, W., Schneider, J., Moosmann, L., Nagl, C.: Representativeness and classification of air quality monitoring stations. Umweltbundesamt Report. (2007)
25. Veness, C.: Calculate distance and bearing between two latitude/longitude points using Haversine formula in javascript. movable type scripts (2011)
26. Luxen, D., Vetter, C.: Real-time routing with OpenStreetMap data. In: 19th ACM SIGSPATIAL International Conference on Advances in Geographic Information System, pp. 513–516. Chicago, USA (2011)
27. Neis, P., Zipf, A. Schmitz, S.: OpenRouteService.org–combining open standards and open geodata. The state of the map. In: 2nd OSM Conference, Limerik, Ireland (2008)

Decision Support System for the Control and Monitoring of Crops

Katty Lagos-Ortiz[1,2(✉)], José Medina-Moreira[1,2], Abel Alarcón-Salvatierra[1], Manuel Flores Morán[1], Javier del Cioppo-Morstadt[2], and Rafael Valencia-García[3]

[1] Facultad de Ciencias Matemáticas y Físicas, Universidad de Guayaquil, Cdla. Universitaria Salvador Allende, Guayaquil, Ecuador
{klagos,aalarcon}@uagraria.edu.ec, {jose.medinamo, manuel.floresmo}@ug.edu.ec
[2] Facultad de Ciencias Agrarias, Universidad Agraria del Ecuador, Av. 25 de Julio, Guayaquil, Ecuador
{klagos,jdelcioppo}@uagraria.edu.ec
[3] Facultad de Informática, Universidad de Murcia, Campus Espinardo, 30100 Murcia, Spain
valencia@um.es

Abstract. One of the main activities in most nations is agriculture and its importance lies in providing food for the growing population. Ecuador has a great natural wealth, it is geographically located on the Equatorial line that gives it its name, which allows it to have a stable climate almost every month of the year with positive consequences for the agricultural sector. The implementation of information and communication technologies (ICTs) in agriculture tends to generate automation and efficiency in processes that were previously carried out manually. Not only does the use of machinery and equipment allow this advance. It is also incorporating computer systems of analysis and help in the decision on fields and crops that allow improving and facilitating productivity, improving land management and your planning. This article presents a decision support system based on expert knowledge of the domain for the control and monitoring of rice, coffee and cocoa crops, which based on information provided by the user and external information such as location and weather It will help in the process of crop selection, control, monitoring, diagnosis, pest prevention, fertilizer selection, among others. These recommendations will be made based on information modeled by experts and other factors in order to reduce costs, increase productivity and optimize the harvest time of the products. The proposed system has been evaluated for the diagnosis of crops affected by diseases, pests and weeds.

Keywords: Decision support · Agriculture · Plague · Insects · Prevention Diagnosis · Crops

R. Valencia-García et al. (Eds.): CITAMA 2019, AISC 901, pp. 20–28, 2019.
https://doi.org/10.1007/978-3-030-10728-4_3

1 Introduction

Agriculture is currently considered one of the most relevant economic, social and environmental activities in almost all countries and its importance lies in providing food for the growing population. The current concern is that the production of food is directly proportional to the growth of population, for this reason it requires more production. Studies point out that the next generation must produce double what is currently being produced. Ecuador has a great natural wealth, it is geographically located on the Equatorial line that gives it its name, which allows it to have a stable climate almost every month of the year with positive consequences for the agricultural sector. Traditionally, the Ecuadorian economy has been based on agriculture, mining and fishing. Since the 1970s, the oil industry has played a vital role in the country development, but from this century onwards, exports of agricultural products are the true engine of economic growth in the country. According to macroeconomic data [1] from January to October 2014, agriculture is the main non-oil export product in Ecuador thanks to different sectors such as bananas, which represent 21.02% of the export of non-oil products, flowers and plants (6.70%), cocoa (5.13%), agro-industry (3.72%) and coffee (1.45%). Adding these three sectors directly related to agriculture, we have 38.08% of total exports of non-oil products.

On the other hand, the agricultural sector involves the greatest workforce with the least investment. However, in this sector there is no equity and its transformation and evolution is a main action that will allow solving unemployment, poverty and inequality in Ecuador.

Nowadays, the implementation of information and communication technologies (ICTs) in agriculture and agrifood chains tends to generate automation and efficiency in processes that were previously carried out manually. Not only does the use of machinery and equipment allow this progress, they are also incorporating computer systems for analysis and decision support on the fields and crops that allow improving and facilitating the productivity of the products to be cultivated, improving land management and use.

This implementation and use of ICTs has led to a new term called e-Agriculture [2] which is a new area that promotes sustainable agriculture and food security through better processes to access and exchange knowledge, through the use of TICs.

Decision Support Systems (DSS) [3] are management information systems that combine analysis models to solve user problems through an interface and allow recommendations for the improvement of those problems. DSS are implemented in areas such as medicine, aero-space, transport, commerce, among others. The development of these DSS is based on the use of Semantic Web technologies, Information and Knowledge Integration, Linked Data, Business Intelligence, among others, which allow to facilitate certain analysis, monitoring and recommendation tasks.

This research project presents the development of a DSS implemented to support farmers and researchers in the identification of crops and their diseases. They also intend to evaluate the decision-making support system based on experts' knowledge that can be implemented in the domain of agriculture. The system will specifically help in the process of crop selection processes prevention, control, monitoring, diagnosis, pest prevention, fertilizer selection, among others. It will take information provided by the user and other external information such as climatology. These recommendations will be made based on information modeled by experts and other factors in order to select crops correctly, increase productivity and optimize the harvest time of the products.

2 Related Works

Liu Chuanju [1] presents that as a unified information system, Internet-based smart decisions approaches which are applied in different field of science. For instance, pattern recognition is used for preventing and controlling pest and diseases in fruit plants; furthermore, smart databases, multimedia and network topologies are implemented for agriculture management. In both cases, the platform implements an identification, analysis and evaluation process to determine their status. By comparing images or recognizing characteristics, the system could provide an idea of what type of pest or diseases are affecting the fruit tree. This information can be evaluated in real time by an expert. Moreover, the system structure is supported by software engineering theory, networking and database technology.

Similarly, H. Kiyoshi [2] argues that in Japan, scientists have developed a system for agriculture information services (FieldTouch). Approximately, 100 people use this technology to improve the agriculture procedures. FieldTouch incorporates multi-scale sensor data for observing the plant growth, as well as offering a recording process for farming procedures in order to support people in a specific decision. Every two weeks, researchers update Rapideye satellite images, meanwhile, from 25 nodes field sensor are updated every 10 min. Furthermore, the National Meteorology Observation Centre provides weather information for the system. A valuable flexible and automated system could be affordable by integrating technology such as AMeDas (Data Source), SOS (Sensor Observation Service) and DSS (Decision Support system) to analyse the behaviour of the climate into agriculture process.

M. Liu [3] states that in specific circumstances, pattern recognition presents some drawbacks because the inaccurate and uncertain information that could be solved by applying a rough set theory provided by mathematics. The evolution of the technology in the field of science, Agricultural Decision Support System (ADSS) presents considerable functions which are applied for many researchers in farmer's activities. Traditional procedures present drawbacks. In this study, a new model of DSS (founded on rough set theory) applies the discernibility matrix to find the main characteristics. Moreover, this research uses an optimized algorithm to attend the incomplete data. In the first stage, the system analyses the information in order to attend the drawbacks. After that, the previously mentioned algorithm is applied to process the incoming data for obtaining the rules. Finally, the ADSS could apply the rough set theory to the process.

In the same way, Phoksawat [4] suggests that Decision Support Systems have been implemented in a considerable number of applications due to their effectiveness. In the agriculture sector, DSS could support the farmers for making decision about complex situations. Pests and plant diseases are the main problems for monoculture activities, affecting the prices. Intercropping is a mechanism to compensate the behavior of the market price and diseases and it has three main goals. The first one is to define the knowledge spectrum of the intercropping planning structure. The second goal is to implement an ontology knowledge-based model, focusing in attending the DSS. The last one is to elaborate an optimization model for intelligent decision making about intercropping systems. In this examination, an ontology knowledge- based and a multi-objective model are used to control the DSS for intercropping structure. This structure relies on a constraint factor that enables each individual to obtain the minimum fee for agriculture while maximizing the incomes.

3 Decision Support System

One of the advantages of the DSS (Decision Support System) is that they are easily adopted by farmers who require information about the care, control and monitoring of their crops. In other words, it helps farmers to face problems with crops, using available data and information provided by experts. This type of systems supports the approach of intelligent agriculture, with the purpose of using the necessary inputs, reducing environmental impacts, optimizing the use of fertilizers and increasing yields, that is, we are looking for ways to use land in an intelligent way.

This article suggests a decision support system that allows farmers through a set of images stored in a database to identify diseases, pests and weeds that affect crops, based on a web and mobile platform. In this way it will be possible to detect the damage caused to their plantations, besides, the system will recommend the type of control and diagnosis using the appropriate fungicides in the right doses. Additionally, the system has a geolocation module which recommends the suitable and recommended crops in that area, crop and harvest times. The information presented by the system has been based on an ontology created with the help of the knowledge that has been obtained from the domain of the experts in the area. With the help of experts, an ontology has been created. The proposed system is a DSS that generates a range of

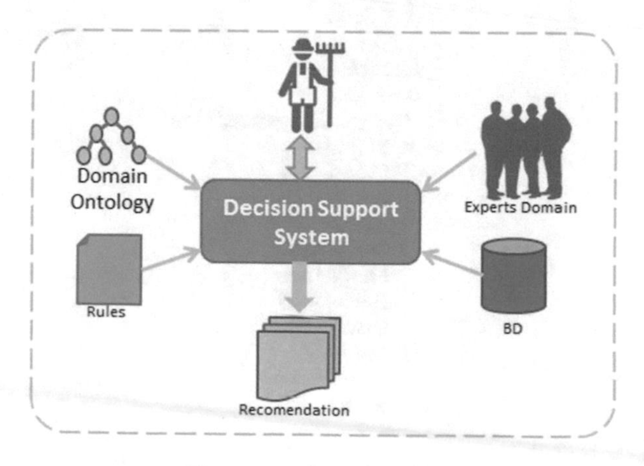

Fig. 1. DSS´s architecture.

recommendations, helping the farmer to make decisions, based on data entered by the user as the required crop (rice, coffee and cocoa) and the problem presented. Its architecture is shown in Fig. 1.

3.1 Crop Control Ontology

Through the design of the ontology for the control of crops, which has been modeled from the knowledge extracted from the domain of experts and using rules of inference and reasoning, it is possible to determine or diagnose the conditions caused in crops by diseases, pests and weeds. Based on these results, we can establish a series of appropriate recommendations to control rice, coffee and cocoa crops. In the creation of ontologies, the use of OWL2 [5], a Web Ontologies language was considered, and Protégé [6] was used as an open source editor to create applications based on ontologies.

The development of the DSS was based on the design of an ontology for which, with the help of the agronomists, the set of diseases, pests and weeds affecting the rice, coffee and cocoa crops was categorized for each crop. It was also established the fungicides used for the treatment of the condition and the dosage recommendation to apply of each fungicide, classified by crops, the concepts and the existing relationships to implement the ontology. Later, existing ontologies were taken as reference in order to design the ontology. Some of them were the Plant Ontology Consortium [7] that offers a vocabulary for the structures of the plants, and those cited in [8]. On the other hand, the group of weeds, pests and diseases affecting the aforementioned crops was clearly specified in such a way that the fungicides that could be used for each condition were grouped together and the relationship between them was established, defining the types, entities and hierarchy of those existing types, properties and relationships.

An extract of the ontology is shown in Fig. 2, which shows the products used to fight phytosanitary problems presented by crops such as diseases, pests and weeds, the recommendations and treatments used, as well as the products and doses for treatments.

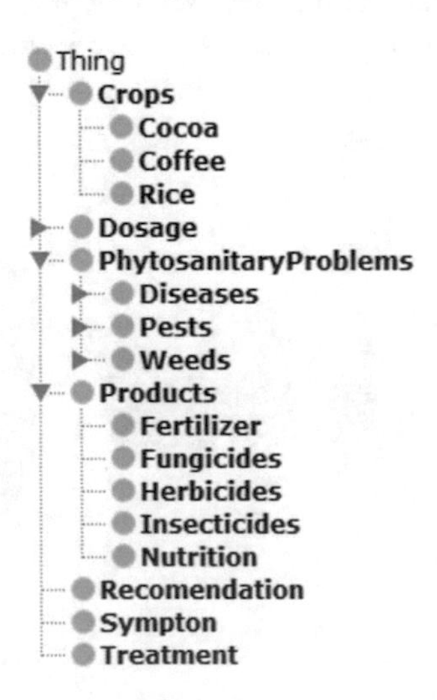

Fig. 2. Ontology for the control of crops.

The ontology defines seven main classes that are: Crops, Doses, Phytosanitary problems, Products, Recommendations, Symptoms and Treatments. They are detailed below:

- Crops. It contains the taxonomy of the cocoa, coffee and rice crops.
- Doses. It contains the amounts of the fungicide and fertilizer products that are applied to the crops.
- Phytosanitary problems. It contains the taxonomy of chemical products: herbicides, fungicides and insecticides that should be placed on diseased crops.
- Products. It lists the taxonomy of the chemicals used in crops.
- Symptoms. It contains the symptoms that – based on experts- the crops present depending on their conditions.
- Recommendation. The experts have established a set of recommendations based on the problems of crops and how to solve them.
- Treatment. It defines the dosage and the way of application of the products to fight the crop affections.

In the ontology domain, you can define some properties to establish the existing relationships between the instances of the classes. Table 1 shows some of these properties.

Table 1. Properties described

Propiedad	Dominio	Rango
has Sympton	Sympton	Diagnosis
has Dosage	Phytosanitary problem	Dosage
has Treatment	Treatment	Diagnosis

3.2 Rule-Based Engine

For the codification of knowledge, we will use the language rules of the Semantic Web (SWRL) [9], which is the programming language used to implement the established rules. To illustrate how the rules are coded, let's take the following rule resulting from the experts' knowledge acquisition phase. Figure 3 shows the rule obtained.

IF	crop is rice
AND	rice has sympton
THEN	recommend diagnostic: PYRICULARIA ORYZAE
AND	recommend treatment: Treatment and products

Fig. 3. Rule sample

The rules created by the authors and that will be used in the system present the following format: $A_1, A_2, \ldots, A_n \rightarrow R$.

The premises A1, A2,…, An, are evaluated and it eventually results in premise R. The rule-based engine comes into play when a farmer enters the desired crop and chooses the crop conditions. It analyzes the input and presents the recommendations about the products that must be applied with the appropriate doses, the rule-based engine presents the treatments and recommendations processed according to the reasoning, so that the end user (farmer) can have options to be able to care for and control their crops.

4 Methodology

This section presents the architecture used in the web system and in the mobile application, which defines an iteration of the front-end with the back-end, that is, the link between the information manager and the information processing, once users make the data income to obtain the desired results in terms of their crops. With the design and implementation of this project, we try to automate certain processes in the agricultural field and allow end users to have a decision support tool that can establish necessary recommendations the moment support is required for monitoring and control of the crops.

4.1 System Architecture

The system is composed of two layers that are the front-end and back-end, terms that refer to the separation between a presentation layer and a data access layer, respectively (see Fig. 4). The presentation layer has been designed in such a way that users can identify and locate the options corresponding to their crops quickly and conveniently. The data access layer is directly linked to the data repository, which has been structured in MySQL, which stores the information related to each of the crops, their diseases, pests and weeds and ways to establish cure mechanisms.

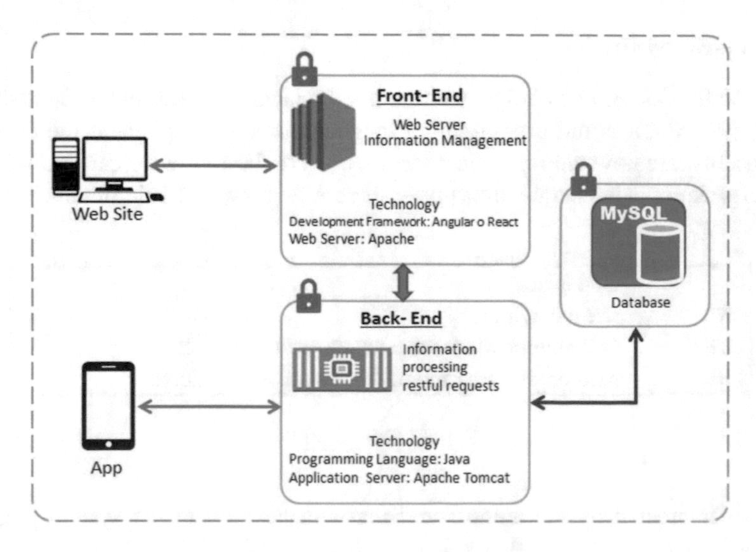

Fig. 4. System architecture

5 Evaluation and Results

A series of experiments has been used to validate the effectiveness of the results of the proposed system. A group of 10 farmers from Costa, a region of Ecuador, have worked with rice, coffee and cocoa crops. Farmers provide information regarding the symptoms of their crops with 10 affections. The farmers' responses served efficiently for the system feedback.

The process was carried out in the following way: (1) The 10 farmers provided input data regarding the affections of their crops, which were chosen through images of affected crops. (2) The purpose of these tests was to determine the treatments and the appropriate product doses for the affected crops. (3) There was also the validation of the results with a set of experts in the field of these crops. (4) As a final step, the effectiveness of the proposal was measured using precision and recovery metrics [10], and the harmonic mean of F-metric [11]. In this measurement, the true positives (TP) are the conditions that the crop associated with a given disease presents. False positives (FP) are symptoms unrelated to a condition and false negative (FN) symptoms mistakenly related. Table 2 shows the evaluation obtained grouped by conditions.

Table 2. Evaluation of results.

Afección	TP	FP	FN	Precision	Recall	F-metric
Pyricularia Oryzae	16	6	5	0,72727	0,76190	0,74419
Sarocladium Oryzae	18	5	4	0,78261	0,81818	0,80000
Hydrellia SP	20	7	6	0,74074	0,76923	0,75472
Tagasodes Orizicolus	18	3	6	0,85714	0,75000	0,80000
Cyperaceas	21	3	6	0,87500	0,77778	0,82353
Average				0,79655	0,77542	0,78449

It can be noted that there are no significant differences between the results. The data presents an average precision of 0.79655, a recovery average of 0.77542 and an average F-metric of 0.78449. The condition that obtained the best precision was Cyperaceas with 0.82353 and is found in rice cultivation.de F-metric de 0,78449.

6 Conclusions and Future Work

The need of technology in the agricultural sector is becoming increasingly important, especially the so called systems of recommendation and decision support (DSS). These systems have become helpful to farmers who need support at the moment of protecting or eliminating any disease that affects their crops.

The evaluation results obtained from the decision support system were positive with F-metric results of 0.78449 for rice crop affections. The development of the DSS contributes with the proposal of the ontology of the diseases that affect the crops as well as the products to control the conditions that help to control the diseases, plagues and weeds; and the set of recommendations and treatments, the rules of inference obtained based on the symptoms of the cultures entered and the corresponding diagnoses.

With the development of this application, it has been demonstrated that it is possible to solve Agro problems through technology using applications that help farmers in making their decisions. We consider future work to include new rules to add the already proposed ones. Additionally, the surveyed farmers have asked for the application to offer other crops than those proposed, to create maintenance modules that allow us to parameterize these requirements and additionally have some image processing technology with which crop affections can be automatically detected.

References

1. Chuanju, L., Ken, C.: Design and implement of web-based intelligent decision support system for prevention and control of fruit tree diseases and pests. In: 2009 4th International Conference on Computer Science and Education, pp. 1269–1271. IEEE (2009)
2. Kiyoshi, H., et al.: FieldTouch: an innovative agriculture decision support service based on multi-scale sensor platform. In: 2014 Annual SRII Global Conference, pp. 228–229. IEEE (2014)
3. Liu, M., Zhang, X., Wang, B.: Research on agricultural decision support system based on rough set theory. In: 2009 International Conference on Future BioMedical Information Engineering (FBIE), pp. 102–109. IEEE (2009)
4. Phoksawat, K., Mahmuddin, M.: Ontology-based knowledge and optimization model for decision support system to intercropping. In: 2016 International Computer Science and Engineering Conference (ICSEC), pp. 1–6. IEEE (2016)
5. Mankovskii, S., Gogolla, M., Urban, S.D., et al.: OWL: Web Ontology language. In: Encyclopedia of Database Systems, Springer, Boston, MA, USA, pp. 2008–2009 (2009)
6. Noy, NF., Crubézy, M., Fergerson, R.W., et al.: Protégé-2000: An open-source Ontology-development and knowledge-acquisition environment. AMIA 2003, Open Source Expo (2003)
7. Plant Ontology Consortium: The Plant Ontology Consortium and plant ontologies. Comp. Funct. Genomics **3**:137–142 (2002). https://doi.org/10.1002/cfg.154
8. Jaiswal, P., Cooper, L., Elser, J.L., et al.: Planteome: A resource for common reference ontologies and applications for plant biology (2016)
9. Horrocks, I., Patel-Schneider, P., Boley, H.: SWRL: A semantic web rule language combining OWL and RuleML. W3C Member (2004)
10. Clarke, S.J., Willett, P.: Estimating the recall performance of Web search engines. Aslib. Proc. **49**:184–189 (1997). https://doi.org/10.1108/eb051463
11. Yang, Y., Liu, X.: A re-examination of text categorization methods. In: Proceedings of 22nd Annual International ACM SIGIR Conference Research and Development in Information Retrieval—SIGIR '99. ACM Press, New York, USA, pp. 42–49 (1999)

A Rule-Based Expert System for Cow Disease Diagnosis

Abel Alarcón-Salvatierra(✉), William Bazán-Vera, Winston Espinoza-Moran, Diego Arcos-Jácome, and Tany Burgos-Herreria

Universidad Agraria del Ecuador, Avenida 25 de Julio, Guayaquil, Ecuador
{jalarcon, wbazan, wespinoza, darcos, tburgos}@uagraria.edu.ec

Abstract. Cow husbandry is one of the main agricultural development sectors in many countries. However, cow diseases problem results in low productivity and restricts the development of this kind of agriculture. Raisers highly depends on a veterinarian to cope with cow disease issues. Unfortunately, there is a lack of veterinarians to serve this sector demands. Therefore, it is necessary to develop innovative solutions focused on solving problems such as the cow disease diagnosis. This work proposes an expert system for cow disease diagnosis. This system allows diagnosing a cow disease based on a set of symptoms provided by the user. For this purpose, the proposed system relies on a set of SWRL-based rules that represent expert knowledge on cow diseases. Our proposal was evaluated by real users from the cow husbandry domain. In this evaluation, the system had to diagnose a cow disease based on a set of symptoms provided by users. The system got promising evaluation results based on the accuracy metric.

Keywords: SWRL rules · Expert system · Cow disease diagnosis

1 Introduction

Cow husbandry is one of the main agricultural development sectors in many countries. This sector helps to meet the domestic demand for meat and dairy products. However, cow diseases problem results in low productivity and restricts the development of this kind of agriculture. Raisers highly depends on a veterinarian to cope with cow disease issues. Unfortunately, there is a lack of veterinarians to serve this sector demands. More specifically, in rural areas, veterinarians are critically needed to provide accessible medical knowledge. Therefore, it is necessary to develop innovative solutions focused on solving problems such as the cow disease diagnosis.

In the last years, information technology has been used in the medical domain. For instance, expert systems are used for animal disease diagnosis and treatments recommendation. An expert system can be defined as a program that attempts to mimic human expertise by applying inference methods to a specific body of knowledge [1]. Such body of knowledge can be modeled through an ontology, which is defined as a formal and explicit specification of a shared conceptualization [2]. In this context,

R. Valencia-García et al. (Eds.): CITAMA 2019, AISC 901, pp. 29–37, 2019.
https://doi.org/10.1007/978-3-030-10728-4_4

SWRL (Semantic Web Rule Language) [3] allow defining rules that can be expressed in terms of OWL concepts to provide deductive reasoning capabilities. Knowledge rules are the prime form for representing human knowledge [4].

The main goal of this research work is to propose a rule-based expert system for cow disease diagnosis. More specifically, this system allows users to diagnose a cow disease based on a set of symptoms provided to the system. This system relies on a set of SWRL-based rules that represent expert knowledge on cow diseases. Also, we aim at testing the effectiveness of the proposed system in terms of accuracy regarding cow disease diagnosis.

The remainder of this paper is structured as follows: Sect. 2 discusses a set of expert systems for disease diagnosis of different animals. Section 3 details the components of the rule-based expert system for cow disease diagnosis proposed in this work, whereas Sect. 4 presents the evaluation performed to test the effectiveness of our proposal in terms of accuracy regarding cow disease diagnosis. Finally, Sect. 5 discusses the research conclusions and future work.

2 Related Work

Thanks to the increasing advances in hardware and software as well as the availability of expert information in the form of animal disease medical records, many researchers have proposed expert systems based on technologies such as ontologies or neural networks to help people in the animal disease diagnosis. For instance, in [5], the authors presented an animal disease diagnosis system that adopts a binary-inference-core mechanism based on two algorithms namely weighted uncertainty reason algorithm and Bayesian method. On the other hand, in [6], the authors presented an Android-based mobile application for cattle diseases diagnosis. This application is composed of an intelligent engine based on the fuzzy neural network that diagnoses cattle diseases and provides raisers with recommendations for tackling the disease. In [7], an expert system for swine disease diagnosis is presented. This system provides user with pictures and description of symptoms to allow users to specify the certainty factor of symptoms.

Some research efforts are focused on the diagnosis of domestic animals' diseases. For instance, in [8], the authors presented an expert system composed of a mobile and Web interfaces through which users provide the pet's symptoms, which are evaluated by the system to infer the disease and to provide information about the symptom-disease relationship. In [9], the authors presented a fuzzy-based expert system for domestic animal disease diagnosis. This system uses a fuzzy logic model to diagnose a disease based on a set of neurological signs.

There are other approaches that focused on fish disease diagnosis. For instance, in [10], a neural network-based expert system for fish disease diagnosis is proposed. This system uses diagnostic medical records of fish diseases to train the neural network. Then, the trained neural network performs the diagnosis of fish disease. In [11], a fish disease diagnosis expert system based on fuzzy neural network is presented. This system deals with the complexity, fuzziness and randomness between symptoms and diseases. On the other hand, in [12], the authors proposed an architecture for a multi-Agent-based fish disease diagnosis expert system. This system uses a task allocation model to perform a cross-regional collaborative diagnosis. Meanwhile, in [13], the authors presented a method for fish disease diagnosis that implements a fuzzy C-means clustering algorithm. The main goal of this method is to achieve a rapid and mass diagnosis of fish diseases. Finally, in [14], the authors presented an SMS (Short Message Service)-based service for fish disease diagnosis. This service performs the diagnosis by using a Bayesian decision-making approach.

In the context of cow disease diagnosis there are few approaches available in the literature. For instance, in [15], the authors proposed a digital management system of cow diseases on dairy farm. This system uses medical records about cow basic information, routine monitoring, and disease prevention. On the other hand, Dairy Cow-vet [16] is a mobile expert system for cow disease diagnosis. The disease diagnosis process implemented by Dairy Cow-Vet uses a significant weight of symptom and the certainty factor of the occurred symptom. Meanwhile, DCDDS [17] is a Web-based system for milch-cow disease diagnosis. This system allows users to select the model to perform the disease diagnosis. The available models are CBR (Case-Based Reasoning), Subjective Bayesian theory and D-S evidential theory. Finally, in [18], the authors presented a system for cow disease diagnosis that employs uncertainty evidence illation to deal with uncertainty. This method has reduced the influence of the subjective factor over the diagnosis accuracy.

As can be observed from the previous analysis, most of the examined frameworks, approaches, or systems focus on the diagnosis of one specific disease or animal. Also, expert systems based on SWRL rules are not specially reported in these works, and only a few proposals focus on cow disease diagnosis. Our proposal addresses these drawbacks through a modular architecture, whose main goal is to allow people in charge of cow husbandry to diagnose the cow disease based on a set of symptoms. To achieve this goal, our proposal uses a set of SWRL-based rules generated from cow disease medical records.

3 A Rule-Based Expert System for Cow Disease Diagnosis

In this section, the components of the rule-based expert system for cow disease diagnosis and their interactions are thoroughly explained and described.

3.1 Architecture and Functionality

The architecture of our proposal has two main components namely Rule-based engine and Knowledge base. These components have different levels of interdependence aiming to allow scalability i.e. to allow generating new rules to better support users on cow disease diagnosis. The functional architecture of our proposal is illustrated in Fig. 1.

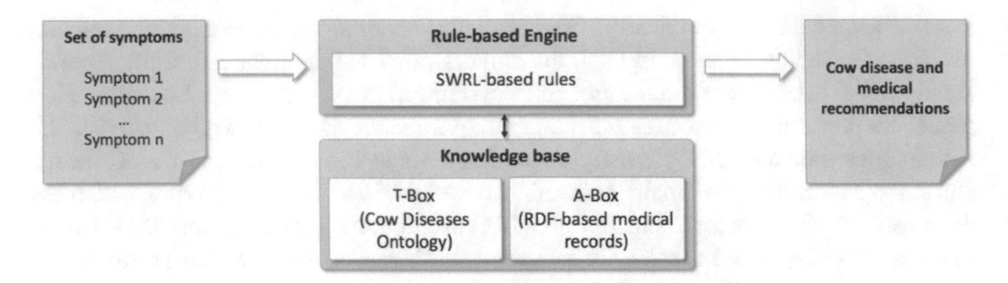

Fig. 1. Architecture of the rule-based expert system proposed in this work.

In a nutshell, the expert system here proposed works as follows. By using this system, users can obtain a cow disease diagnosis, by providing a set of symptoms. For instance, let us suppose that user provide cow symptoms such as vomiting, depressed appetite, among others. Then, the rule-based engine diagnoses the cow disease based on the rules previously generated from a set of medical records of cow disease diagnosis and treatments. Once the cow disease is diagnosed, the system provides a set of medical recommendations to deal with the diagnoses disease. The functions of the main components of our proposal are thoroughly described below.

3.2 Knowledge Base

This knowledge base consists of a T-Box and an A-Box. The T-Box layer, also known as ontology, describes the structure of classes, relationships and constraints of the knowledge base. The A-Box layer consists in the knowledge bases instances. The goal of this repository is to allow using reasoning to infer new knowledge, in this case, to diagnose a cow disease by means of a rule-based engine, which is described in Sect. 3.3. The T-Box and A-Box layers are described below:

- Ontology. This repository includes an ontology for cow diseases classification including a classification of symptoms and their relationships. This ontology was designed in conjunction with a group of researchers from the Agrarian University of Ecuador. This ontology was designed and implemented by using the OWL2 language [19] following the knowledge engineering methodology proposed in [20], which consists of six phases: problem assessment, data and knowledge acquisition, prototype development, complete system development, system evaluation, and integration and maintenance of the system. For the purposes of this work, only first five phases were performed. The ontology considers concepts such as:
  - Disease. This class contains a taxonomy of cow diseases. The cow diseases considered by this class are bluetongue, babesiosis, ketosis, bronchial pneumonia, and BVD (Bovine Viral Diarrhea).
  - Symptom. This class refers a taxonomy of symptoms, which are defined as phenome accompanying something, in this case, a cow disease, and is regarded as evidence of its existence. Some of the symptoms described by this ontology are reduced feed intake, tremor, dehydration, edema, vomiting, depressed appetite, dark urine, milk loss, among others.

  - Body system. This class contains a taxonomy of body systems where symptoms occur. Some of the body system considered in this work are nervous, respiratory, digestive, reproductive, urinary, endocrine, muscular, circulatory, among others.
  - Treatment. This class refers to a treatment, which is defined as an effort to cure or improve a disease or other health problem.
- A-Box. It contains an RDF-based repository of medical records for the diagnosis and treatment of cow diseases. This repository was developed from a set of 200 health records in veterinary medicine, which include the monitoring of different vital and physical parameters of cows during disease diagnosis and treatment.

3.3 Rule-Based Engine

As was previously mentioned, knowledge rule is the prime form to represent human knowledge. Considering this fact, the expert system described in this work adopts a rule-based approach to model the relations between cow diseases and their corresponding symptoms. In other words, this approach allows our proposal to simulate experts' reasoning process by using a set of rules.

The rules defined in this proposal were defined through the SWRL language. All rules defined follow the format presented in Eq. 1

$$R_1, R_2, R_3, \ldots, R_n \rightarrow D \tag{1}$$

Where *Ri, R2, .., Rn* are atomic formulas depicting conditions and *D* is the cow disease diagnosed when the conditions are fulfilled. For instance, the definition of the Bluetongue disease is represented as follows

```
disease(?x) ∧ hasSymptom(?x, "laminitis") ∧ hasSymptom(?x,
"abortion") ∧ hasSymptom(?x, "dyspnea") ∧ hasSymptom(?x,
"weight loss") ∧ hasSymptom(?x, "milk loss") ∧ hasSymptom(?x,
"mouth congestion") ∧ hasSymptom(?x, "diarrhea") → isDisease
(?x, "bluetongue")
```

The above rule specifies that the Bluetongue disease will be inferred when property *hasSymptom* matches with the symptoms that define the disease, in this case, "laminitis", "abortion", "dyspnea", "weight loss", "milk loss", "mouth congestion", and "diarrhea". Finally, it should be mentioned that once disease is diagnosed, the proposed system provides users with a set of medical recommendation that help them to treat such disease. All this information is contained in the knowledge base described in the Sect. 3.2.

4 Evaluation

In the context of expert systems for cow disease diagnosis, there are no standard datasets that can be easily exploited to perform an evaluation. However, in this work we have adopted an approach that evaluate the system in terms of accuracy regarding cow disease diagnosis. The following sections describe the evaluation method and discuss the obtained results.

4.1 Method

To evaluate our proposal, we relied on a group of 20 farmers from the Costa Region of Ecuador, ranged from 30 to 48 years old, with experience in cow husbandry. This evaluation consisted of four phases which aimed to achieve the goal established. These phases are described below:

1. Our proposal was introduced to the participants. Then, farmers involved in this evaluation process were asked to provide a set of symptoms they perceive when a specific cow disease is presented. The cow diseases considered were those ones described by the ontology presented in Sect. 3.2 namely bluetongue, babesiosis, ketosis, bronchial pneumonia, and BVD (Bovine Viral Diarrhea). As result of this phase, we obtained 125 sets of symptoms, specifically 25 sets for each cow disease considered in this process.
2. Each set of symptoms generated in phase 1 was provided as input to the rule-based engine to get a diagnosis.
3. All disease diagnosis provided by our proposal were evaluated by a group of veterinarians with experience in cow husbandry, more specifically, in the diagnosis and treatment of cow diseases. In other words, this group of experts determined if the diagnosis provided by our proposal was correct.
4. To evaluate the effectiveness of our proposal in terms of accuracy regarding cow disease diagnosis we employ the accuracy metric whose formula is shown in Eq. 2.

$$Accuracy = C/A * 100 \tag{2}$$

Where C refers to the cases where cow disease was correctly diagnosed by our proposal. Meanwhile, A refers to all cow disease diagnosis cases that were performed by the system. The evaluation results are presented and discussed in the following section.

4.2 Results

Table 1 presents the evaluation results regarding cow disease diagnosis. As can be observed, the disease diagnosis accuracy rate was high for all cow diseases. More specifically, the disease diagnosis accuracy rate of our proposal ranged from 0.76 to 0.84, with a mean accuracy of 0.808. The proposed expert system got the highest accuracy rate to the Bluetongue disease, with a rate of 0.88. Meanwhile, our proposal got the lowest accuracy rate to the Ketosis and BVD (Bovine Viral Diarrhea) diseases, with a rate of 0.76.

As can be observed in Table 1, there are no significant differences among the results obtained by our proposal for each disease considered. This fact can be interpreted as a good effectiveness of this system for diagnosing cow diseases. In general, cow disease diagnosis cases where a small set of symptoms was provided showed lower accuracy rates than cases where a big set of symptoms was provided. This is understandable, since a wider set of symptoms helps to better diagnose the disease.

Furthermore, the diseases that were not correctly diagnosed have symptoms that are difficult to perceive by people. Unfortunately, this affects the disease diagnosis process.

Table 1. Evaluation results in terms of accuracy regarding cow disease diagnosis.

Cow disease	Accuracy
Bluetongue	0.88
Babesiosis	0.84
Ketosis	0.76
Bronchial pneumonia	0.80
BVD (Bovine Viral Diarrhea)	0.76
Avg.	0.808

In conclusion, although our proposal correctly diagnosed most cases used in this work, the evaluation revealed that this system needs to improve its way of dealing with diseases that have several symptoms in common. That is, this system must include new weighting mechanisms to appropriately diagnoses a disease when the symptoms provided by the user match two or more cow diseases.

5 Conclusions

This paper presented an intelligent system for cow disease diagnosis and treatments recommendation. The main component of this system is a rule-based engine implemented using the SWRL language. The system was evaluated by using real-world users. Despite the evaluation results are promising, we are aware that our proposal could be improved by including weighting mechanisms that improve disease diagnosis when symptoms match two or more diseases. Also, we plan to develop new rules that consider cow living environment variables such as weather, water, grass, among others. This information could help to better diagnose the disease as well as to determine the cause of the it, thus helping farmers to select the right medicine for treatment. It must be remarked that the rule-based engine works for only five cow diseases. As future work, we are planning to generate new rules to deal with other cow diseases. For this purpose, we are planning to develop a Web application through which veterinarians update the symptom-disease relationships as well as to provide treatments to deal with the diseases. Finally, it must be mentioned that another contribution of our work is the ontology-based knowledge base for cow disease diagnosis and treatment, which represents a source of computable knowledge that can be exploited by other researchers. In this sense, we plan to add more medical records that help us to generate more and better rules to improve cow disease diagnosis.

References

1. Darlington, K.: The essence of expert systems. Prentice Hall, Upper Saddle River (2000)
2. Studer, R., Benjamins, V.R., Fensel, D.: Knowledge engineering: Principles and methods. Data Knowl. Eng. **25**, 161–197 (1998)
3. Horrocks, I., Patel-Schneider, P.F., Boley, H., Tabet, S., Grosof, B., Dean, M., others: SWRL: a semantic web rule language combining OWL and RuleML. W3C Memb. Submiss. **21**, 79 (2004)
4. Wenxue, T., Xiping, W., Jinju, X.: An animal disease diagnosis system based on the architecture of binary-inference-core. In: 2010 IEEE Fifth International Conference on Bio-Inspired Computing: Theories and Applications (BIC-TA), pp. 851–855. IEEE (2010)
5. Wenxue Tan, Xiping Wang, Jinju Xi: An animal disease diagnosis system based on the architecture of binary-inference-core. In: 2010 IEEE Fifth International Conference on Bio-Inspired Computing: Theories and Applications (BIC-TA), pp. 851–855. IEEE (2010)
6. Anggraeni, W., Muklason, A., Ashari, A.F., Wahyu, A., Darminto: developing mobile intelligent system for cattle disease diagnosis and first aid action suggestion. In: Proceedings - 2013 7th International Conference on Complex, Intelligent, and Software Intensive Systems, CISIS 2013, pp. 117–121. IEEE (2013)
7. Nusai, C., Cheechang, S.: Uncertain knowledge representation and inferential strategy in the expert system of swine disease diagnosis. In: 2014 International Conference on Information Science, Electronics and Electrical Engineering, pp. 1872–1876. IEEE (2014)
8. Bilen, M., Isik, A.H., Yigit, T.: Expert system software for domestic animals. In: 2017 International Conference on Computer Science and Engineering (UBMK), pp. 130–134. IEEE (2017)
9. Jampour, M., Jampour, M., Ashourzadeh, M., Yaghoobi, M.: A fuzzy expert system to diagnose diseases with neurological signs in domestic animal. In: 2011 Eighth International Conference on Information Technology: New Generations, pp. 1021–1024. IEEE (2011)
10. Deng, C., Wang, W., Gu, J., Cao, X., Ye, C.: Research of fish disease diagnosis expert system based on artificial neural networks. In: Proceedings of 2013 IEEE International Conference on Service Operations and Logistics, and Informatics, pp. 591–595. IEEE (2013)
11. Gu, J., Deng, C., Lin, X., Yu, D.: Expert system for fish disease diagnosis based on fuzzy neural network. In: 2012 Third International Conference on Intelligent Control and Information Processing, pp. 146–149. IEEE (2012)
12. Ma, D., Chen, M.: Building of an Architecture for the fish disease diagnosis expert system based on multi-agent. In: 2012 Third Global Congress on Intelligent Systems, pp. 15–18. IEEE (2012)
13. Miao-Jun, X., Jian-Ke, Z., Hui, L.: A method for fish diseases diagnosis based on rough set and FCM clustering algorithm. In: 2013 Third International Conference on Intelligent System Design and Engineering Applications, pp. 99–103. IEEE (2013)
14. Miaojun, X., Jianke, Z., Xiaoqiu, T.: Intelligent fish disease diagnostic system based on SMS platform. In: 2013 Third International Conference on Intelligent System Design and Engineering Applications, pp. 897–900. IEEE (2013)
15. Li, L., Wang, H., Yang, Y., He, J., Dong, J., Fan, H.: A digital management system of cow diseases on dairy farm. Presented at the October 22 (2011)
16. Nusai, C., Chankeaw, W., Sangkaew, B.: Dairy cow-vet: A mobile expert system for disease diagnosis of dairy cow. In: 2015 IEEE/SICE International Symposium on System Integration (SII), pp. 690–695. IEEE (2015)

17. Rong, L., Li, D.: A web based expert system for milch cow disease diagnosis system in China. Computer and Computing Technologies in Agriculture, vol. II, pp. 1441–1445. Springer US, Boston (2007)
18. Zhang, Y., Xiao, J., Fan, F., Wang, H.: The expert system of cow disease diagnosis basing on the uncertainty evidence illation. In: 2010 4th International Conference on Bioinformatics and Biomedical Engineering, pp. 1–4. IEEE (2010)
19. Grau, B.C., Horrocks, I., Motik, B., Parsia, B., Patel-Schneider, P., Sattler, U.: OWL 2: the next step for OWL. Web Semant. Sci. Serv. Agents World Wide Web. **6**, 309–322 (2008)
20. Negnevitsky, M.: Artificial intelligence : a guide to intelligent systems. Addison-Wesley Boston (2005)

Internet of Things and Monitoring Systems

On-Line Monitoring of Bioelectricity from a Microbial Fuel Cell Using Fishery-Industry Wastewater

Carlos Banchón[1(✉)], Catherine Peralta[1], Tamara Borodulina[1], Maritza Aguirre-Munizaga[2], and Néstor Vera-Lucio[2]

[1] Faculty of Agricultural Sciences, School of Environmental Engineering, Universidad Agraria del Ecuador, Av. 25 de Julio y Pio Jaramillo, P.O. BOX 09-04-100, Guayaquil, Ecuador
{cbanchon, tborodulina}@uagraria.edu.ec
[2] School of Computer Engineering, Faculty of Agricultural Sciences, Universidad Agraria del Ecuador, Av. 25 de Julio y Pio Jaramillo, P.O. BOX 09-04-100, Guayaquil, Ecuador
{maguirre, nvera}@uagraria.edu.ec

Abstract. The fishery industry accounts negative environmental impacts to natural resources because of its wastewater with high amount of organic matter. The present study is focused on the production of bioelectricity using a single chamber of Microbial Fuel Cell (MFC), as well as the removal of suspended solids from a fishery industry wastewater. For this, an Arduino microcontroller and voltage sensors were used to monitor the bioelectricity production. The voltage peaked almost 750 mV in four cells at the end of 48 h. The removal of suspended solids reached almost a 87%. The harnessing of energy and wastewater treatment using a MFC is a promising method for valorization of wastes in the fishery industry.

Keywords: Arduino · Automation · Turbidity · Degradation · Electrochemical microorganisms

1 Introduction

Almost a 90 per cent of all wastewater in developing countries is discharged untreated directly into rivers, lakes or oceans [1]. For example, industrial wastewaters from fisheries are a severe concern because they generate saline wastewaters rich in proteinic nitrogen, organic matter and salts [2, 3]. To treat these wastewaters, most traditional treatment processes consume energy. For instance, treatment of organic-rich wastewater consumes about 3% (1.05×10^{10} W) of all electrical power produced in the USA each year [4]. However, it is estimated that wastewaters contain approximately 9.3 times more energy than what is currently needed for its treatment in a modern municipal wastewater treatment plant [5]. The energy locked in wastewater is mainly present in three forms: organic matter (~1.79 kWh/m^3); nutritional elements such as nitrogen, phosphorous (~0.7 kWh/m^3); and thermal energy (~7 kWh/m^3) [6].

R. Valencia-García et al. (Eds.): CITAMA 2019, AISC 901, pp. 41–48, 2019.
https://doi.org/10.1007/978-3-030-10728-4_5

A way to capture the energy available in wastewaters is through a Microbial Fuel Cell (MFC). A MFC can generate electricity from nearly all sources of biodegradable organic matter from food industries wastewaters [7–9]. MFCs use microorganisms like *Geobacter sulfurreducens, Shewanella putrefaciens, Desulfovibrio desulfuricans, Geothrix fermentans* and *Rhodopseudomonas palustris* to convert the chemical energy of biomass into electricity [10].

Given that bioelectricity production is a form to prevent the contamination of natural resources from the fishery industry, the contributions in the present study are:

- The removal of turbidity and total solids in wastewater by a MFC single chamber.
- An on-line system using Arduino to monitor the production of bioelectricity in mV.

2 Related Works

A microbial fuel cell functions as a bio-electrochemical transducer that convert microbial reducing power (generated by the metabolism of organic substrates) into electrical energy [11]. Some bacteria can transfer electrons outside the cell to a terminal electron acceptor (for example, a metal oxide like iron oxide) [5]. These bacteria that can exogenously transfer electrons, called *exoelectrogens*, can be used to produce electricity in an MFC [4]. Oxygen should be avoided because it will inhibit the electricity generation. The chamber where no oxygen is present, the anode chamber, is where bacteria grow; and the cathode chamber, is where the electrons react with the cathode under the presence of oxygen [10, 12]. The two electrodes are connected with a wire, which is connected with the voltage sensor.

The present project is based on the smart voltage and current monitoring system (SVCMS), in which a three phase electrical system is monitored [13]. The voltage was read using an Arduino platform as a microcontroller, and then the data is sent wirelessly to an Android application. This kind of voltage measurement using a microcontroller was developed in an urine powered battery energy harvesting system [14]. There, the aim was to collect energy from human liquid waste. A MFC with a brush-shaped anode and a cathode of graphite was able to generate electrical power (9.2 mW m^{-2}) and to degrade 80% of organic matter from domestic wastewater [15]. In the coal industry, coking wastewater was used for bioelectricity generation. The removal of carbon and nitrogen from coking wastewater reached almost 82%, while producing a power of 538 mW m^{-2} [16].

The importance of maintaining sensors that allow the transmission of voltage data in a wireless manner is highlighted, as is the case of the design, manufacture and testing of low-cost sensors that can detect bacteria at a limit of detection of only 10 cells mL^{-1} [17]. As another example, an Arduino system was implemented to monitor the alcohol production in a distillation column [18].

Thus, the implementation of this monitoring system using Arduino have been used in academia mainly to give a technological support of low cost and easy access to industry. In previous studies [19], we demonstrate that Arduino platforms are an excellent pedagogic tool easy to access and synchronize.

3 Methodology

3.1 Microbial Fuel Cell

The MFC was inoculated with wastewater from a biological treatment plant of an Ecuadorian fishery industry. Microorganisms were isolated from the fishery wastewater using nutrient broth and cultivated a 30 °C (Merck, Germany). The MFC was operated in a 500 mL batch cycle, at approx. 25 °C. Hence, a wastewater residence time of 1 week was achieved. The cathode was constructed from stainless steel mesh (0.3 mm diameter); it was covered with an activated carbon film. Copper wires were connected to the voltage sensors as showed in Fig. 1. Each voltage sensor maintains a positive and a negative pole, which are connected directly to the voltage transmitters.

Fig. 1. Working prototype of a MFC

3.2 Bioelectricity On-Line Monitoring

The on-line monitoring system worked with an Arduino microcontroller [20], and voltage sensors to monitor four MFC chambers [13, 21]. The sensors sent information to a wireless card. The data is stored on the web[1]. The real-time data transmission is shown through an interface designed in php [22]. This programming language is related to a database in MySQL which allows the information to be stored every time the cells are connected to the prototype.

Figure 2 shows the model implemented to carry out voltage measurement through the cells. In this case, the MFCs were directly connected to a voltage sensor that maintained a range of 24.45–25000 mV. The data was monitored every 5 min.

[1] http://meteorologiauae.uagraria.edu.ec/proyecto20/.

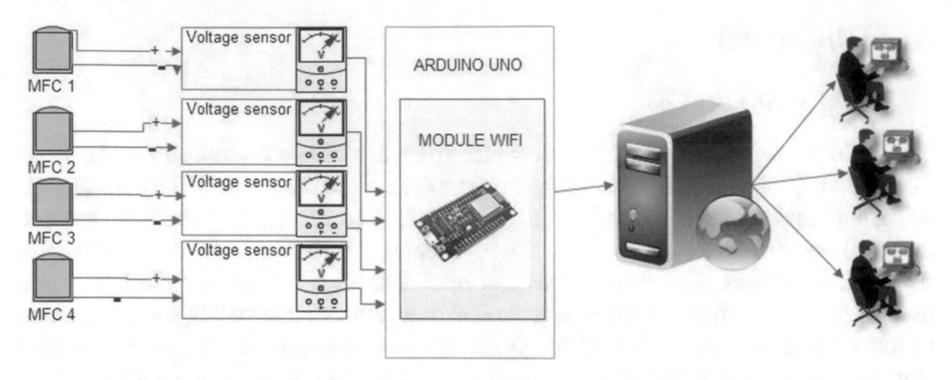

Fig. 2. Schematic presentation of a data transmission model for bioelectricity monitoring of four MFCs.

Figure 3 shows the connections in the Arduino microcontroller. The sensors are connected to the analogous pins of the Arduino. This connection enables that the prototype works like an analog multimeter. The Arduino code converted the analog data captured by the sensors into millivolts (mV). The standard values of the resistance were included in the design of the voltage sensors, as well as the maximum factor of Arduino reception; in this case, this is 5000 mV, which was received as a numerical factor of 1024. As a limitation of the present study, problems were detected in the adaptation of the device depending on the security of the network where the readings were activated, which was solved by assigning a static IP to the Nodemcu module [23].

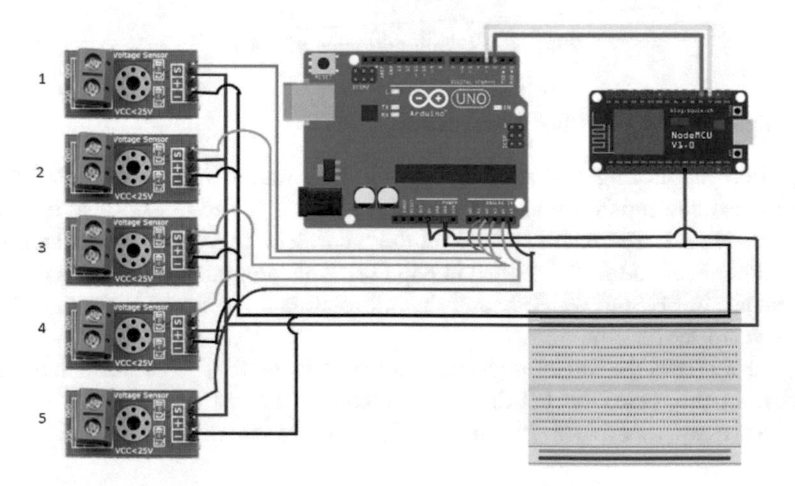

Fig. 3. Schematic presentation of a Arduino prototype

```
float R1 = 30000;
float R2 = 7500;
float reading1 = analogRead(sensorvol1);
```

```
float compute1 = (reading1 * 5.0) /1024.0;
float compute1 = compute1 /(R2/(R1 + R2));
float compute1 = (compute1*1000);
```

Figure 4 shows the most important interfaces designed in the web page for the control of the values measured by the prototype; in this case, three functions are highlighted in the web page: (i) Data download, in which the user of the laboratory can download a. csv file containing all the monitored results; (ii) Statistics, which helps to verify through a descriptive statistic the changes produced in the voltage levels of the cells that are connected to the prototype; (iii) Real time function, which is used to control the voltage of the cells from any browser and is updated every five minutes.

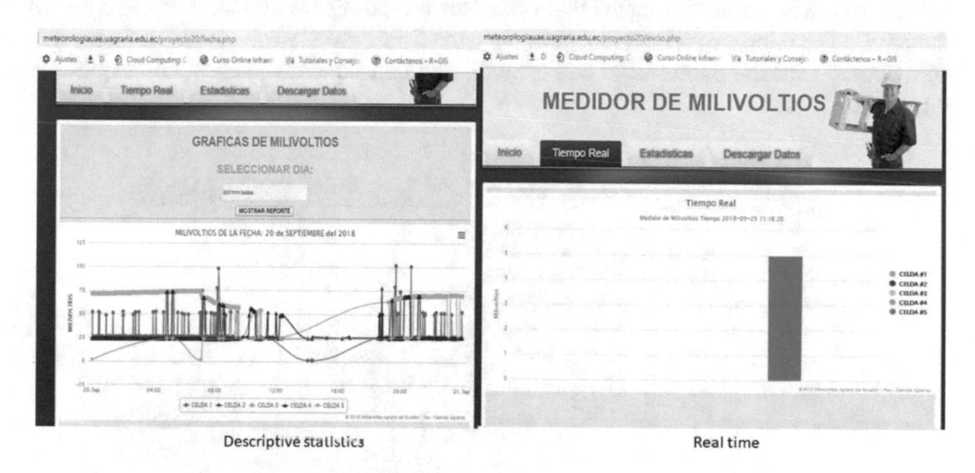

Fig. 4. Web interface.

4 Results and Discussion

Regarding the implementation of the prototype as an embedded system [24], the (i) utility, (ii) complexity of use and its installation, and (iii) intuitivity were evaluated from the users (researchers), as shown in Table 1. According to the survey, an 80% of users answered that the application is very comfortable for measuring voltage variations from the MFCs; also, the application does not need to be installed by the researchers, since the interface can be used and reviewed from any computer with internet access through a browser. On average, 60% of the researchers who used the system considered that the interface had a high degree of intuitivity.

Table 1. Results of a survey to researchers about the embedded system.

Factor	Not comfortable at all		Not so comfortable		Somewhat comfortable		Very comfortable		Extremely comfortable		Users
Utility	0	0%	0	0%	0	0%	4	80%	1	20%	5
Installation	0	0%	0	0%	0	0%	0	0%	5	100%	5
Intuitivity	0	0%	0	0%	0	0%	2	40%	3	60%	5
Mean	0	0%	0	0%	0	0%	2	40%	3	60%	5

In a MFC, most anaerobic microorganisms have the potential of being bioelectricity catalysts. If final electron acceptors such as oxygen, nitrate and sulfate are absent in the anode chamber, therefore a proper electron shuttling is performed [4]. Figure 3 shows the time profile of four MFCs recorded during three days. The medium was not replaced with fresh wastewater in order to determine the maximum voltage production from a 500 mL wastewater sample. The peak voltage was 750 mV at the end of 48 h. This is a voltage that related works have obtained as well, which demonstrates a successful growth of electrochemically active microorganisms [16, 25].

According to Fig. 5, the increase in cell voltage up to 750 mV suggests that anaerobic microorganisms were metabolic active. According to Table 2, in this case, the fishery wastewater electrical conductivity (1320 µS cm^{-1}) is similar to groundwater; thus, it is a relevant factor that promotes the prompt conduction of electrons in the MFC. The microorganisms isolation from the fishery wastewater is also a factor that contributed with the exponential cell growth, and consequently with the bioelectricity production.

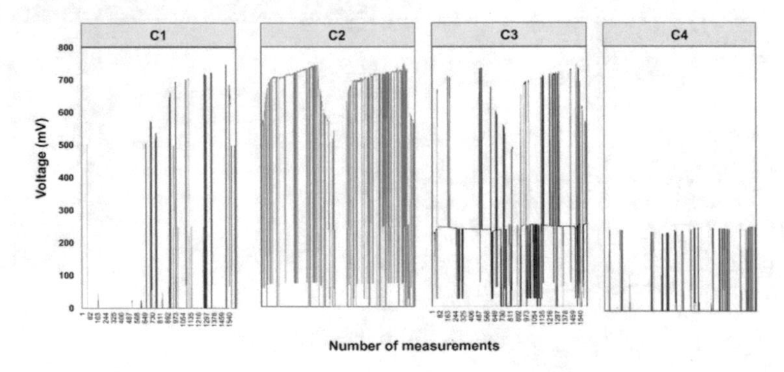

Fig. 5. Voltage generation in four MFCs during three days

The performance of the MFC was also evaluated in terms of solids removal efficiency. According to Table 2, the initial turbidity was removed by almost 87% (MFCs 1 to 3). The dissolved solids were not removed at all. In this case, dissolved solids increased due to the microbial exponential growth which tend to form microflocs in wastewater. As consequence of the microflocs formation, although the turbidity was mostly removed, the total solids were removed only by approx. 20%. This means that, the wastewater should be treated by another operation to remove dissolved solids in order to get completely clarified. The performances of solids degradation were attributed to an anaerobic-aerobic environment, in which suspended organic matter is also mostly removed.

Table 2. Summary of water parameters before and after the operation of four MFCs

Parameter	WW	MFC1	MFC2	MFC3	MFC4
pH	6.5	6.6	6.6	6.8	6.7
Temperature (°C)	25.4	23.9	23.9	24.3	24.5
Dissolved solids (mg/L)	955	1064	1142	1091	1038
Electrical conductivity (μS/cm)	1320	1470	1570	1500	1430
Turbidity (NTU)	342	73*	29*	63*	154
Total solids (mg/L)	512	552	408*	2556	372

WW = Wastewater, C1–C4 = MFCs

*Statistical significance ($p < 0.05$)

5 Conclusions and Future Work

In the present study, a one chamber activated carbon and stainless steel cathode MFC showed the feasibility of producing bioelectricity from fishery industry wastewater. The bioelectricity production in four independent MFCs reached a peak of 750 mV, and they treated wastewater from fisheries by the removal of solids up to 87% of turbidity. As a limitation of the present study, problems were detected in the adaptation of the device depending on the security of the network where the readings were activated, which was solved by assigning a static IP to the Nodemcu module[2].

In perspective, we project to monitor temperature profiles in order to correlate them with the voltage production, as well as configuring the prototype as a datalogger. Moreover, due to the increasing number of reports about electrochemically active microorganisms, a genetic characterization of most microorganisms from fishery wastewaters is relevant to enhance bioelectricity production.

Acknowledgements. The authors are grateful to Christian Chavez Vergara for his kind help on the implementation of the MFCs.

References

1. Corcoran, E., Nellemann, C., Baker, E., Bos, R., Osborn, D., Savelli, H.: Sick water: the central role of wastewater management in sustainable development: a rapid response assessment. In: UNEP/GRID-Arendal, Arendal, Norway (2010)
2. Lefebvre, O., Moletta, R.: Treatment of organic pollution in industrial saline wastewater: a literature review. Water Res. **40**, 3671–3682 (2006)
3. You, S.J., Zhang, J.N., Yuan, Y.X., Ren, N.Q., Wang, X.H.: Development of microbial fuel cell with anoxic/oxic design for treatment of saline seafood wastewater and biological electricity generation. J. Chem. Technol. Biotechnol. **85**, 1077–1083 (2010)
4. Logan, B.E.: Exoelectrogenic bacteria that power microbial fuel cells. Nat. Rev. Microbiol. **7**, 375–381 (2009)

[2] http://nodemcu.com/index_en.html.

5. Liu, W., Cheng, S.: Microbial fuel cells for energy production from wastewaters: the way toward practical application. J. Zhejiang Univ. Sci. A **15**, 841–861 (2014)
6. Gude, V.G.: Wastewater treatment in microbial fuel cells - an overview. J. Clean. Prod. **122**, 287–307 (2016)
7. Borole, A.P.: Microbial fuel cells and microbial electrolyzers. Interface Mag. **24**, 55–59 (2015)
8. Das, D.: Microbial Fuel Cell: A Bioelectrochemical System that Converts Waste to Watts. Springer International Publishing, New Dehli (2017)
9. Wen, Q., Wu, Y., Zhao, L., Sun, Q.: Production of electricity from the treatment of continuous brewery wastewater using a microbial fuel cell. Fuel **89**, 1381–1385 (2010)
10. Franks, A.E., Nevin, K.P.: Microbial fuel cells, a current review. Energies **3**, 899–919 (2010)
11. Ieropoulos, I.A., Greenman, J., Melhuish, C., Hart, J.: Comparative study of three types of microbial fuel cell. Enzyme Microb. Technol. **37**, 238–245 (2005)
12. Scott, K., et al.: Microbial Electrochemical and Fuel Cells: Fundamentals and applications (2016)
13. Mnati, M., Van den Bossche, A., Chisab, R., Mnati, M.J., Van den Bossche, A., Chisab, R. F.: A smart voltage and current monitoring system for three phase inverters using an android smartphone application. Sensors **17**, 872 (2017)
14. Saini, S.S., Singh, J., Bhatia, H., Sidhu, E.: Dead cell extracted-urine powered battery energy harvesting system. In: 2016 IEEE 7Th Power India International Conference (2016)
15. Buitrón, G., Cervantes-Astorga, C.: Performance evaluation of a low-cost microbial fuel cell using municipal wastewater. Water Air Soil Pollut. **224**, 1470 (2013)
16. Huang, L., Yang, X., Quan, X., Chen, J., Yang, F.: Amicrobial fuel cell-electro-oxidation system for coking wastewater treatment and bioelectricity generation. J. Chem. Technol. Biotechnol. **85**, 621–627 (2010)
17. Jiang, J., et al.: Smartphone based portable bacteria pre-concentrating microfluidic sensor and impedance sensing system. Sensors Actuators B Chem. **193**, 653–659 (2014)
18. Barcia-Quimí, A.F., León-Munizaga, N.C., Aguirre-Munizaga, M., Hernandez, L., Vergara, V.: Automation of a distillation column of packed bed for an alcohol solution using arduino. Rev. Int. Investig. y Docencia. **2**, 1–7 (2017)
19. Gómez-Chabla, R., Aguirre-Munizaga, M., Samaniego-Cobo, T., Choez, J., Vera-Lucio, N.: A Reference Framework for Empowering the Creation of Projects with Arduino in the Ecuadorian Universities (2017)
20. Arduino: Arduino SwitchCase. https://www.arduino.cc/
21. Gasperi, M., Hurbain, P.: "Philo": Voltage Sensors. In: Extreme NXT: Extending the LEGO MINDSTORMS NXT to the Next Level, pp. 119–126. Apress, Berkeley, CA (2009)
22. Gilmore, W.J.: Beginning PHP and MySQL. Apress, Berkeley, CA (2010)
23. NodeMcu – An open-source firmware based on ESP8266 wifi-soc
24. Mulfari, D., Celesti, A., Fazio, M., Villari, M., Puliafito, A.: Using embedded systems to spread assistive technology on multiple devices in smart environments. In: Proceedings - 2014 IEEE International Conference on Bioinforma. Biomed. IEEE BIBM 2014, pp. 5–11 (2014)
25. Kim, J., et al.: Effects of various pretreatments for enhanced anaerobic digestion with waste activated sludge. J. Biosci. Bioeng. **95**, 271–275 (2003)

A Monitoring System for Lettuce Cultivation in an NFT Hydroponic System: A Case Study

Raquel Gómez-Chabla(✉), Karina Real-Avilés, Kléber Calle, César Morán, Freddy Gavilánez, Diego Arcos-Jácome, and Cristhian Chávez

Universidad Agraria del Ecuador, Av. 25 de Julio y Pio Jaramillo, P.O. BOX 09-04-100 Guayaquil, Ecuador
{rgomez,kreal,kcalle,cmoran,fgavilanez, darcos}@uagraria.edu.ec, cristhian.chavez1994@gmail.com

Abstract. Agriculture is an activity that is an essential part of the world economy. Due to climate change and the need for food self-sufficiency, innovative solutions to improve production conditions are needed. Hydroponics is an alternative for food self-sufficiency because it allows increasing the yield, growth, quality of different crops. In this sense, there are works focused on the automation of hydroponic systems to monitor the environment and control the nutrient solution for the optimal development of plants. However, in Ecuador, there are no reports of monitoring systems for hydroponic systems that allow reducing the losses of plants in crops. This document describes a case study of the implementation of a monitoring system for lettuce cultivation in an NFT (Nutrient Film Technique) hydroponic system whose objective is to improve the quality and quantity of food. This system provides a mobile application that allows monitoring control variables to take corrective actions and regulate different environmental factors. The system was evaluated in a real scenario obtaining more precise values than traditional methods.

Keywords: Hydroponic · NFT · Lettuce · Monitoring system · Sensors

1 Introduction

Agriculture is an activity that is an essential part of the world economy [1]. Due to climate change and the need for food self-sufficiency, innovative solutions to improve production conditions are needed [2, 3]. Nowadays, the gap between agricultural producers and technology has been reduced thanks to systems such as remote sensing systems, geographic information systems, as well as institutions where expert people exchange information, which in turn allows generating agriculture models centered on the content [4]. Hydroponics is an alternative for food self-sufficiency because it allows increasing the yield, growth, quality of different crops [5]. Hydroponics is based on the fact that the root system of plants is not established in the soil but in the nutritive solution that contains the necessary elements for the growth of these ones. Hydroponic

R. Valencia-García et al. (Eds.): CITAMA 2019, AISC 901, pp. 49–58, 2019.
https://doi.org/10.1007/978-3-030-10728-4_6

systems can be used for the cultivation of the following species: Solanaceae (tomato, chili, eggplant, potato), Liliaceous (onion, garlic, chives, leeks), Cruciferous (turnip, cabbage, cauliflower, broccoli, watercress), cucurbits (cucumber, squash, melon, watermelon), Umbelliferous (cilantro, celery, parsley and carrot), and compound (lettuce). Figure 1 presents a classification of techniques for hydroponic systems. This classification is based on the medium used for the development of the plant roots. The most used techniques are the substrate with drip irrigation, floating root, and culture by NFT (Nutrient Film Technique). These techniques consider parameters such as luminosity, pH, electrical conductivity, the distance of sowing, and oxygenation of the nutritive solution.

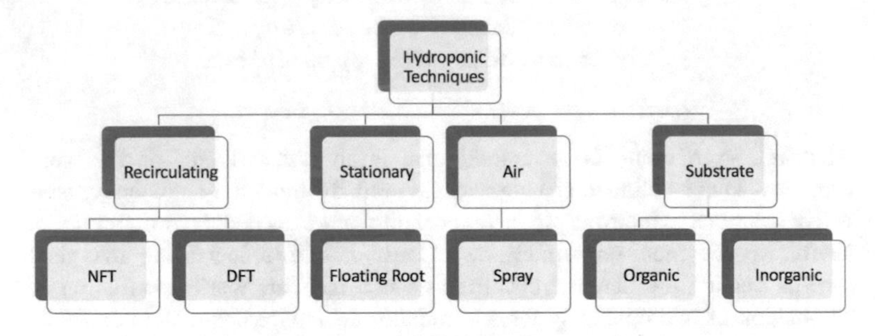

Fig. 1. Hydroponic techniques classification.

Table 1 shows the concentration of cations and anions that a nutrient solution must have for the optimal development of plants. On the one hand, the optimal concentration of cations in the solution is 45% of two calcium cations (Ca++), 35% of a potassium cation (K+), and 20% of two magnesium cations (Mg++). On the other hand, the required anion concentration is 60% of a nitrate anion (NO3-), 5% of a diacid phosphate anion, and 35% of two sulfate anions (SO4–).

Table 1. Optimal concentration of cation and anion in a nutrient solution.

Cation	Concentration	Anion	Concentration
Ca++	45%	NO3-	60%
K+	35%	H2PO4-	5%
Mg++	20%	SO4–	35%

The pH should be kept within the range of 5.2 to 6.3 in ornamental plants, and 5.5 to 6.5 in vegetables. A pH below 5.0 can cause deficiencies of Nitrogen (N), Potassium (K), Calcium (Ca), Magnesium (Mg), Boron (B) mainly. Meanwhile, a pH above 6.5 can decrease the assimilation of Phosphorus (P), Iron (Fe), Manganese (Mn), Zinc (Zn) and Copper (Cu). Finally, the electrical conductivity (EC) allows monitoring of the salt content of the nutrient solution. A high level of EC indicates a high concentration of salts. However, regulated maintenance of a nutrient solution for most species requires that this parameter is maintained in the range 1–2 μS/m.

Hydroponics has advantages such as the reuse of water, the ease of controlling external factors, and the reduction of traditional agricultural practices. However, this technique also has disadvantages such as high installation cost, rapid propagation of pathogens, and the need for specialized knowledge in the management of this kind of crops [6]. In this sense, there are works focused on the automation of hydroponic systems [7, 8] to monitor the environment and control the nutrient solution for the optimal development of plants. In Ecuador, there are no reports of monitoring systems for hydroponic systems that allow reducing the losses of plants in crops. This document describes a case study of the implementation of a monitoring system for lettuce cultivation in an NFT (Nutrient Film Technique) hydroponic system [9] whose objective is to improve the quality and quantity of food. This system uses an Open Source architecture [10] for the development of a mobile application [11] that allows monitoring control variables to take corrective actions and regulate different environmental factors.

2 Related Works

In countries such as Ecuador, Argentina, Bolivia, Peru, Uruguay and Venezuela, simplified hydroponics [12] has been implemented as a productive tool for food security. This technique allows growing vegetables all year round in places such as patios and terraces. In addition, it provides employment opportunities and family integration. There is a great dependence between hydroponics and agriculture [13]. For instance, in [14], the use of the hydroponic technique for the cultivation of five species is evaluated. The results indicate that this technique works best for green forage cultivation. The growth and future of hydroponics in Latin America [15] will depend to a large extent on the development and adaptation of less sophisticated commercial systems. In 2013, the FAO (Food and Agriculture Organization of the United Nations) defined four priority areas: (1) strengthen public policies to increase productivity in a sustainable manner; (2) guarantee food sovereignty; (3) strengthen institutional and legal frameworks for the management of food safety and quality; and (4) contribute to the consolidation of public environmental policy through the conservation, valuation and sustainable management of biodiversity and natural resources. These priorities can be achieved through the application of innovative technologies that demand little investment. For instance, there is a great advance in technologies for the development of monitoring systems such as sensors [16]. These devices provide large amounts of environmental information that can be used to improve crop production, growth, and health [17]. In addition, sensors are used to improve the environmental impact and efficiency of agriculture [18, 19].

3 Monitoring System for Lettuce Cultivation in an NFT Hydroponic System

The monitoring system for lettuce cultivation in an NFT hydroponic system use three main technologies namely Arduino, Android, and MySQL. Arduino is an open-source electronics platform based on easy-to-use hardware and software. The system implements an Arduino-based device that allows obtaining information of the environment in real time through programmable sensors. Some of these parameters are environmental temperature, relative humidity, pH, electrical conductivity and dissolved oxygen in the water. Also, the system provides an Android-based mobile application that allows users to visualize data collected by the Arduino-based monitoring system. Finally, the system implements a MySQL-based database to store all data collected by the sensors.

3.1 Case Study

The system here described uses an NFT-based system for lettuce cultivation. NFT requires a nutrient solution that is prepared and stored in a tank. This solution is driven by an electric pump to the culture channels. Figure 2 depicts the structure of the NFT hydroponic system for lettuce cultivation.

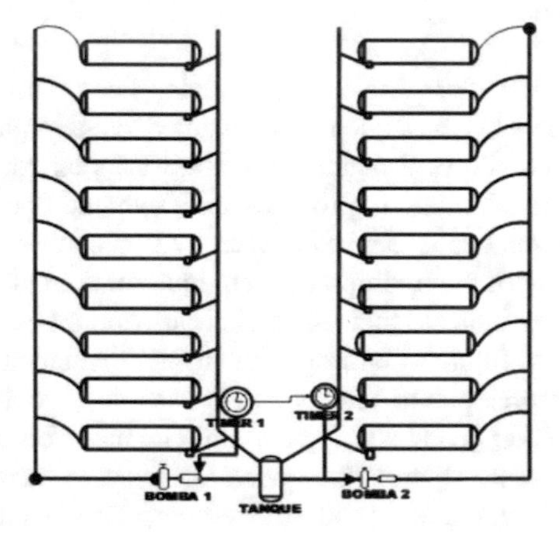

Fig. 2. Structure of the NFT hydroponic system for lettuce cultivation.

The hydroponic system uses recirculation times of 8 (right side) and 6 hours (left side) This system allows the cultivation of 198 Green Salad Bowl lettuce plants whose maturing time is 60 to 65 days. The hydroponic system combines the laminar flow of nutrients with the floating root system which consists of two modules with an independent pump and nine PVC culture channels. These cultivation channels lead the nutritive solution of the tank to the upper part of the channels. This allows to properly supply oxygen, water, and nutrients for the growth of the plant.

During the case of the study described in this paper, daily manual measurements of pH [20], electrical conductivity, humidity, and temperature levels were performed. The recirculation of the nutrient solution by means of timers for the activation of the electric pumps was carried out manually. In addition, manual corrections of pH and EC levels were performed, as well as additions of nutrient solution to the tank to ensure optimal levels of dissolved oxygen. The proposed system collects automatically data on the aforementioned parameters and stores the information in the database. Based on this information, the system generates notifications in real time to the producer about the variations that occur during the crop cycle, thus allowing users to perform the corrective actions.

3.2 Monitoring System Design

Figure 3 depicts the block diagram of the proposed monitoring system. This diagram represents the communication between the sensors and the Arduino board. The input data is the data captured by the sensors. Specifically, the sensors used are HiLetgo DHT22/AM2302 for humidity, PT-1000 Temperature Kit Atlas Scientific for temperature, pH Kit Atlas Scientific for pH, Dissolved Oxygen Kit Atlas Scientific for O2, and Conductivity K 1.0 Kit Atlas Scientific for EC.

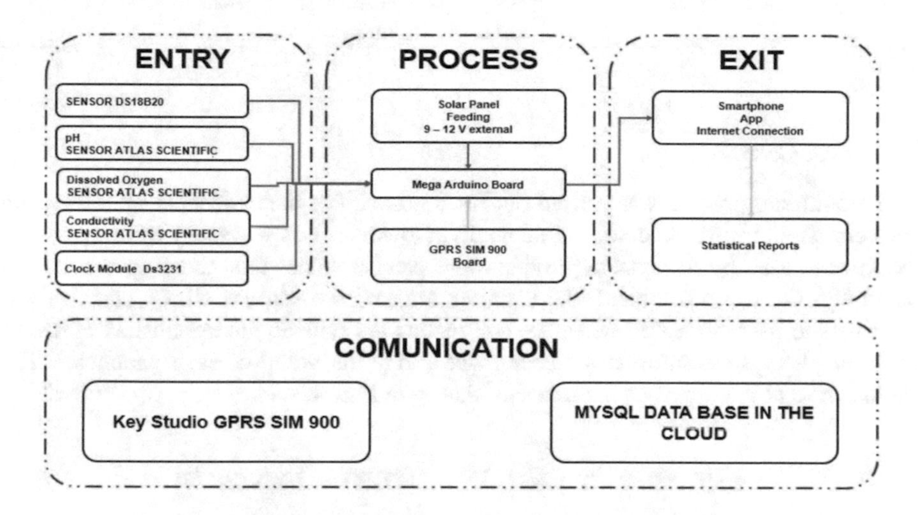

Fig. 3. Block diagram of the monitoring system.

Table 2 presents the optimal ranges of each of the parameters considered by the system. These ranges aim to allow a successful cultivation of the lettuce. The data collection process is done with Arduino Mega and the information collected is stored in the cloud using the GPRS SIM 900 board [21, 22]. Finally, the user can obtain reports of this information through the mobile application. Daily crop monitoring is important since each parameter establishes an indicator of various aspects to be considered for the correct cultivation process. For example, pH must be controlled to keep nutrient elements available in the solution. The EC indicates the content of total salts in the solution (a high level of EC indicates a high concentration of salts and vice versa). The presence of oxygen in the nutrient solution is necessary for the development of the plant and the growth of the roots. In this sense, the proposed system generates alerts, when the values obtained are outside the optimum range.

Table 2. Optimum range of parameters for cultivation of lettuce.

Parameter	Sensor	Range alert	Below range	Above range
Humidity	HiLetgo DHT22/AM2302	70%–80%	Wrong value. Check the device	Add nebulized water
Temperature	PT-1000 Temperature Kit Atlas Scientific	20ºC–24ºC	Wrong value. Check the device	Add nebulized water
pH	pH Kit Atlas Scientific	5.5–6.5 RTA	Provide more nutrient solution	Adjust the solution with phosphoric acid
O2	Dissolved Oxygen Kit Atlas Scientific	8–9 mg O2/lt	Increase daily recirculation	Reduce daily recirculation
EC	Conductivity K 1.0 Kit Atlas Scientific	500 to 800 uS/m	Add more concentrated solution	Add water to dilute the salt

3.3 Test Case

The monitoring process was carried out for 33 days. The data collected by the system are very similar to those collected manually. However, it is necessary to mention that the sensors used by the system provide more precise values than those ones collected manually. On the other hand, the programming of the sensors allows sending text messages to users to notify when the parameters are outside the established range. In addition, the system informs the user twice a day the value of each parameter. The alerts arrive at the email of the users as shown in Fig. 4.

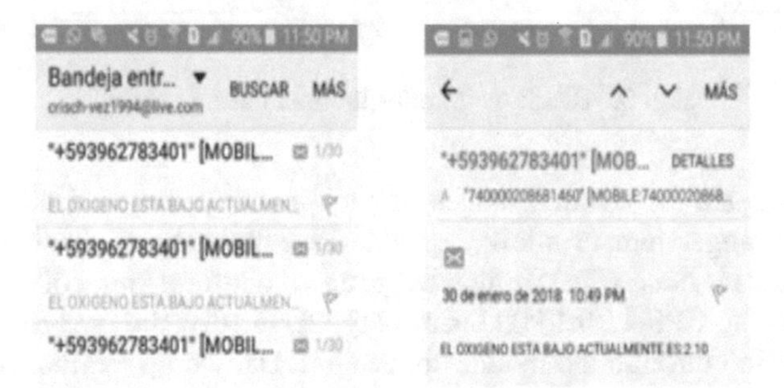

Fig. 4. Alerts generated by the monitoring system.

The data and messages are transmitted to a MySQL database hosted in the cloud. This process is performed by means of a SIM900 GPRS board and a mobile phone chip. The Arduino Mega board is powered by a solar panel battery. In addition, this board processes the data that is displayed through the mobile application and that serves as a basis for the generation of alerts.

4 Results

This section presents a comparison of the values of each parameter obtained manually and through the monitoring system described in this work. For the purposes of this process, the nutritional needs of the lettuce were calculated and the recirculation times of the sheet, temperature, pH, electrical conductivity and dissolved oxygen in the nutrient solution were analyzed. Table 3 presents the values obtained automatically and manually. As can be seen, the average variation between the values obtained is minimal, being more accurate the data obtained by the sensors of the system. The parameters of pH and temperature had an average variation of 0.1 and the HR parameter showed no variation. Furthermore, there is a great variation between the manual and automatic reading of the EC and O2 parameters.

Table 3. Parameters

Parameter	Manual	Automatic
pH	6.7	6.8
CE (uS/m)	610.7	604.9
O2 (mg/Lt)	3.9	6.1
HR (%)	75.9	75.9
T (ºC)	25	24.9

Figure 5 presents a graph with the number of times the system detected out-of-range values compared to traditional methods. It can be seen that the proposed system detected more times out-of-range values for the pH and EC parameters. With respect to the O2 parameter, the system did not detect out of range values that were detected manually. Finally, the manual method and the proposed system detected the same number of out-of-range values for the pH and temperature parameters.

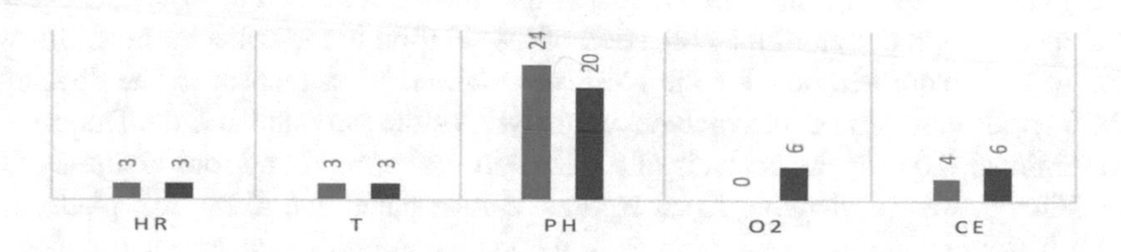

Fig. 5. Out-of-range values detection

Finally, it is worth mentioning that the system generates statistical reports like the one shown in Fig. 6. These reports can be exported to Excel for further analysis.

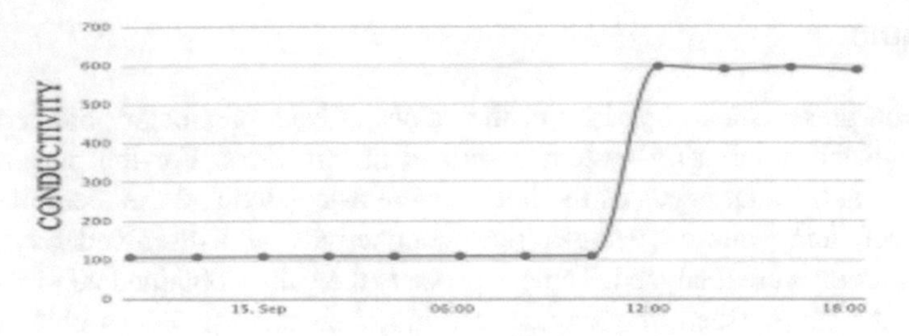

Fig. 6. Statistical report of the electric conductivity.

5 Conclusions and Future Work

Hydroponics allows the rational use of nutrients, water, space, time and labor. In addition, it allows to reduce the energy consumption and the attack of pathogens in the raw material. From the technological point of view, the system allows to control the constant recirculation of water and essential nutrients that allow the plant to grow without stress. From the evaluation described in this work, it was observed that the data collected by the proposed system are more precise than those ones obtained by traditional methods. Furthermore, the system detected more times the out-of-range values, which allowed the user to be alerted through the mobile application. These alerts allow users to take the necessary actions to improve the process of lettuce cultivation. Based on the results obtained, the system could improve the lettuce cultivation process since it allows maintaining the proper nutrients in the water and prevents the appearance of diseases by alerting the user when there are values out of range in the parameters considered. These alerts allow the user to carry out the necessary corrective actions quickly. Finally, it is important to mention that this system uses innovative technologies such as mobile applications, GPRS boards, solar panels and other devices based on the IoT [23, 24].

As future work, it is recommended to use a UV plastic cover to avoid the passage of sun rays harmful to the crop. The drainage pipes must have a sufficient height to cause turbulence at the moment of the collection of the solution in the collector tank, this will allow to gain more aeration for the nutritive solution. With respect to the developed mobile application, this could include an analysis of the variance and the Duncan test with Alpha of 0.05 for the analysis of root length, leaf weight and root biomass of the crop. Finally, we are planning to incorporate a solar panel and a free IoT platform to have a history of the data and apply various prediction and management techniques.

References

1. van der Ploeg, J.D.: Peasant-driven agricultural growth and food sovereignty. J. Peasant. Stud. **41**, 999–1030 (2014). https://doi.org/10.1080/03066150.2013.876997
2. Zabel, F., Putzenlechner, B., Mauser, W.: Global agricultural land resources - a high resolution suitability evaluation and its perspectives until 2100 under climate change conditions. PLoS One (2014). https://doi.org/10.1371/journal.pone.0107522

3. Tittonell, P.: Ecological intensification of agriculture-sustainable by nature. Curr. Opin. Environ. Sustain. **8**, 53–61 (2014). https://doi.org/10.1016/j.cosust.2014.08.006
4. Phupattanasin, P., Tong, S.-R.: Applying information-centric networking in today's agriculture. APCBEE Proc. **8**, 184–188 (2014). https://doi.org/10.1016/j.apcbee.2014.03.024
5. Wortman, S.E.: Crop physiological response to nutrient solution electrical conductivity and pH in an ebb-and-flow hydroponic system. Sci. Hortic. (Amsterdam) **194**, 34–42 (2015). https://doi.org/10.1016/j.scienta.2015.07.045
6. Lee, S., Lee, J.: Beneficial bacteria and fungi in hydroponic systems: Types and characteristics of hydroponic food production methods. Sci. Hortic. (Amsterdam) **195**, 206–215 (2015). https://doi.org/10.1016/j.scienta.2015.09.011
7. Budye, D., Dhanawade, P., Kirti, P., Mahesh, P., Gupte, A.: Automation in hydroponic system. Int. Jounral Res. Eng. Appl. Manag. **3** (2018)
8. Atmadja, W., Liawatimena, S., Lukas, J., Nata, E.P.L., Alexander, I.: Hydroponic system design with real time OS based on ARM cortex-M microcontroller. IOP Conf. Ser. Earth. Environ. Sci. (2018). https://doi.org/10.1088/1755-1315/109/1/012017
9. Nguyen, N.T., McInturf, S.A., Mendoza-Cózatl, D.G.: Hydroponics: a versatile system to study nutrient allocation and plant responses to nutrient availability and exposure to toxic elements. J. Vis. Exp. (2016). https://doi.org/10.3791/54317
10. Gómez-Chabla, R., Aguirre-Munizaga, M., Samaniego-Cobo, T., Choez, J., Vera-Lucio, N.: A Reference framework for empowering the creation of projects with arduino in the ecuadorian universities. Commun. Comput. Inf. Sci. (2017). https://doi.org/10.1007/978-3-319-67283-0_18
11. Viseur, R.: From open source software to open source hardware. **378**, 286–291 (2012). https://doi.org/10.1007/978-3-642-33442-9_23
12. Izquierdo, J.: Simplified hydroponics: a tool for food security in Latin America and the Caribbean. Acta Hortic. 67–74 (2007)
13. Goddek, S., et al. Navigating towards decoupled aquaponic systems: a system dynamics design approach. Water (Switzerland) (2016). https://doi.org/10.3390/w8070303
14. Al-Karaki, G.N., Al-Hashimi, M.: Green fodder production and water use efficiency of some forage crops under hydroponic conditions. ISRN Agron **2012**, 1–5 (2012). https://doi.org/10.5402/2012/924672
15. Rodríguez-Delfín, A: Advances of hydroponics in Latin America. Acta Hortic. 23–32 (2012)
16. Patil, P., Kakade, S., Kantale, S., Shinde, D.: Automation in hydroponic system using PLC. Int. J. Sci. Technol. Adv. **2**, 69–71 (2016)
17. Wu, H., Aoki, A., Arimoto, T., Nakano, T., Ohnuki, H., Murata, M., Ren, H., Endo, H.: Fish stress become visible: a new attempt to use biosensor for real-time monitoring fish stress. Biosens. Bioelectron. **67**, 503–510 (2015). https://doi.org/10.1016/j.bios.2014.09.015
18. Montoya, A.P., Obando, F.A., Morales, J.G., Vargas, G.: Automatic aeroponic irrigation system based on Arduino's platform. J. Phys. Conf. Ser. (2017). https://doi.org/10.1088/1742-6596/850/1/012003
19. Siregar, S., Sari, M.I., Jauhari, R.: Automation system hydroponic using smart solar power plant unit. J. Teknol. **78**, 55–60 (2016). https://doi.org/10.11113/jt.v78.8713
20. Cambra, C., Sendra, S., Lloret, J., Lacuesta, R.: Smart system for bicarbonate control in irrigation for hydroponic precision farming. Sensors (Switzerland) (2018). https://doi.org/10.3390/s18051333
21. Costanzo, A.: An arduino based system provided with GPS/GPRS shield for real time monitoring of traffic flows. In: AICT 2013 - 7th International Conference on Application of Information and Communication Technologies, Conference Proceedings (2013). https://doi.org/10.1109/icaict.2013.6722710

22. Szydlo, T., Nawrocki, P., Brzoza-Woch, R., Zielinski, K.: Power aware MOM for telemetry-oriented applications using GPRS-enabled embedded devices – levee monitoring use case. 2014 Federated Conference on Computer Science Information System, FedCSIS 2014 **2**:1059–1064 (2014). https://doi.org/10.15439/2014f252
23. Sosa, E.O., Alberto, D., Internet, G., Tecnol, R.C.: Internet del futuro. Rev. Cienc. y Tecnol, Desafíos y perspectivas (2014)
24. González, D.R.: Arquitectura y Gestión de la IoT. Rev. Telem@tica 12 (2013)

Automated Hydroponic Modular System

Erick González-Linch[1], José Medina-Moreira[1], Abel Alarcón-Salvatierra[1], Silvia Medina-Anchundia[1], and Katty Lagos-Ortiz[1,2](✉)

[1] Facultad de Ciencias Matemáticas y Físicas, Cdla, Universidad de Guayaquil, Universitaria "Salvador Allende", Guayaquil, Ecuador
{erick.gonzalezli, jose.medinamo, abel.alarcons}@ug.edu.ec, klagos@uagraria.edu.ec

[2] Facultad de Ciencias Agrarias, Universidad Agraria del Ecuador, Av. 25 de Julio, Guayaquil, Ecuador

Abstract. The purpose of this document is to evaluate the implementation of technology to the hydroponics technique. Knowing that classical cultivation requires the permanent care of the human being for the irrigation of water and tender of the nutritive solution, it is proposed to use technology to replace the basic functions of crop irrigation through the use of programmable and reprogrammable microcontrollers, thus following the free hardware culture, this paper evaluates a scalable system with sensors, an open source hardware and software platform to interact between analog and digital signals. Within the steps that comprehend the realization of the project, the methodology based on components, provides the appropriate guidance for the analysis and choice of hardware, thus allowing the integration of the electronic components to work along with the appropriate coding; as a result the project will collect the pH value present in the nutrient solution, if the value is not adequate, it will automatically activate the insertion of the solution to the container reservoir, to subsequently operate the automated water irrigation for hydroponic farming. Another relevant aspect is that the project can be improved with other devices that will allow an even greater control of variables and data of hydroponics crops.

Keywords: Hydroponics · Automated · Modular · Capsicum annuum

1 Introduction

These days, worldwide, we can see an incensement in the food needs of people, with the demographic explosion and the higher consumption in homes, it is necessary to look for more sources of food; this situation is increasingly aggravated by the erosion of soils, which leads the search for new areas of agricultural production that are naturally becoming scarce. This situation is harder in developing countries where agriculture is one of the main sources of work and income of its population, Ecuador

R. Valencia-García et al. (Eds.): CITAMA 2019, AISC 901, pp. 59–67, 2019.
https://doi.org/10.1007/978-3-030-10728-4_7

being a mainly agricultural country,[1] necessarily must look for new ways to cultivate the land and one of them are hydroponic crops [1, 2].

Hydroponics or hydroponic agriculture is a technique that consists in cultivating without soil, also called agriculture without soil. In different parts of the world, this type of crop is widely used, especially in places where conventional agriculture presents adverse conditions. In Ecuador, acceptance is being reached for this type of crops where they can become a good alternative for food in homes [3].

However, although the benefits of this type of agriculture are encouraging, performing these types of crops at a commercial level is often complicated given the amount of nutrients that the water must have so that it can nourish the plants in an adequate manner; Due to this, we developed a MODULAR HYDROPONIC SYSTEM controlled by a microprocessor [4]. The use of this technology will allow us to manage in a better way this type of crops, since with the help of sensors it will be possible to monitor and control the level of hydrogen concentration present in some solutions.

When sensing all pH levels in the water, the system will have a module that will proceed to inject saline solution to level the necessary values so that the crops can continue their growth stage.

2 Related Works

Japiest [5] is an intelligent integral system for the prevention, diagnosis and control of diseases affecting tomato (Solanum lycopersicum), which are grown in hydroponic greenhouses, in which temperature, humidity and nutrient consumption can influence the diseases or pests that crops may get, this tool helps farmers to make decisions to apply appropriate control treatments to treat a disease or pest.

Domingues et al. [6] presents the development of a software-managed system that monitors the conductivity and the PH of the lettuce (Lactuca sativa), as it is one of the highest consumption vegetables, the study was carried out for 24 h, meaning the production cycle, it is also possible to adjust any variation, by using solenoid valves that provide acid or nutrient solutions. The proposal was evaluated in two different ways: one was hydroponics in a controlled greenhouse with the system devices and those cultivated manually in the soil. As results, it was obtained that using the developed system analyzing agronomic and chemical parameters, the precocity in the harvest was demonstrated, offering the farmer an effective and viable alternative.

Ibayashi et al. [7] in his paper A Reliable Wireless Control System for Tomato Hydroponics, proposes a reliable wireless control system for a tomato (Solanum lycopersicum) hydroponics crop, this system has fault tolerance and self-healing functions to recover from packet transmission failures. The system achieved a hydroponic solution supply control in real time without data loss by using a 400 MHz WSN band in the tomato hydroponic harvest.

[1] http://www.ecuadorencifras.gob.ec/estadisticas-agropecuarias-2/.

Under the approach presented in this article, Saaid [8] presents an investigation in which the pH level in the crop´s water solution will be maintained automatically by the microcontroller and measured by a sensor, in such a way that the pH level begins to change, and the effects of the water solution were determined. This research also focuses on the ability to adjust the pH value in the water solution for (DWC) Deep Water Culture that allows the normal and more efficient growth of the plant.

In his work Ferentinos [9], develops two separate fault detection models: one for the detection of faulty operation of a deep-trough hydroponic system which is caused by mechanical, actuator or sensor faults, and one for the detection of a category of biological faults [10], this work supports in the previous analysis for the first model, where he properly sensed a faulty situation in a very short time, within 20 or 40 min for the devices and shows that a neural network have useful generalization capabilities.

As Savvas [11] this development was applied to a pepper (Capsicum annuum) crop in closed hydroponic systems in a glasshouse, with two different irrigation regimes by adding NaCI, also combined with two different levels of irrigation frequency in a two-factorial experimental design. After nearly three months they converged gradually to maximal levels depending on the NaCl treatment, which is a similar approach as Tuberosa [12].

3 Methodology

For the development of this hydroponic modular system controlled by microprocessors, it was sought to relate the areas of agriculture with technology; and with this, achieve to encourage the community to be able to implement this system for the improvement of the techniques used in hydroponic crops [13].

For the implementation of the system, the component-based software process model was used; which is based on the reuse of a large number of components. Component-based projects seek re-use informally, since when assigning a work team to a project, they develop the skill and expertise of codes, similar to the one requested and many times even the designated project, since when searching and using it only is modified according to the needs of our system, making its reuse independent of the processes used for its development and thus providing specific functionalities.

The fundamental part that was used for the execution of this system was the free hardware, which allows us to implement the system without a significant increase in costs. This hardware has been implemented using microcontrollers and one of the most popular is the so-called ARDUINO [14, 15], which is based on open source hardware and software.

It details the way in which this electronic device was used, where signals captured by sensors were entered, after developing algorithms for the management of these inputs and thereby generating output signals that are used to indicate the status of the hydrogen concentration in the water of the hydroponic crops and in turn, if necessary, saline solution is added to obtain the required pH levels [16].

The system does not use a “software” as such, the configuration of the inputs and outputs is made directly using the Arduino integrated development environment (IDE) which is open-source.

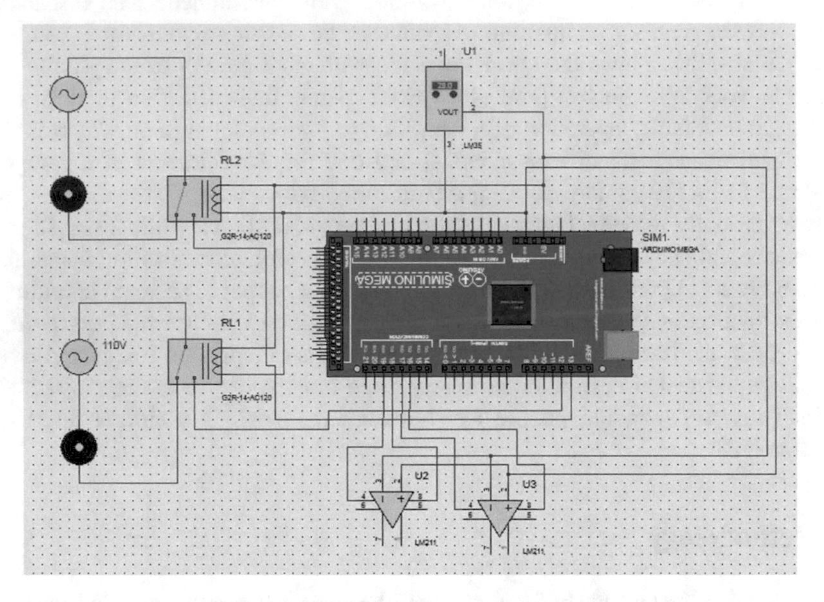

Fig. 1. Hardware connection diagram

Fig. 2. Actual hardware connected

Figure 1 shows the hardware connection diagram, where the connections are established between an Arduino Mega 2560 card, as the general controller of the system, an 8-channel relay, that will allow to operate the valves in charge of the automated irrigation of the crop and several different sensors that will measure the different analogue values of the harvest and command actions deployed by the Arduino, as for example irrigation of the substrate in case it is necessary. It is observed that this is a simple circuit that could be implemented in an agile and simple way, and does not need a large space to be installed, it is completed by a stainless steel structure and ordinary PVC pipes to contain the resulting crops (Fig. 2).

A. Sensors

To obtain data of the level of hydrogen concentration in water, it is necessary to use sensors, for the development a pH sensor was used, its operation consists of detecting the electric current, which can be a negative or positive value, and its generated by the existence of hydrogen ions present in the water. This device is composed of an electrode, a BNC connector and a sensor circuit.

The sensor circuit is an integrated element that facilitates reading the pH level, it uses an asynchronous communication protocol, used in computer systems due to its compatibility with other components.

In addition to the measurements of the concentration of hydrogen in the hydroponic harvest, its humidity must also be monitored, this action consists of measuring the humidity due to the variation of its conductivity and thus be able to activate the irrigation system.

The humidity sensor used is the FC-28, which is a plate that allows us to obtain the measurement as an analogue value or as an digital output signal activated when humidity exceeds a certain threshold, which are values ranging from 0 (submerged in water) to 1023 (in the air or dry substrate).

As for the digital output, this is triggered by exceeding the humidity threshold that is adjusted by a potentiometer built into the YL-83 module, to obtain a LOW signal, if the substrate is not wet and HIGH when the humidity exceeds the set value, it will depend on the type of soil or substrate and the presence of fertilizers, since different plants do not require the same humidity, so the calibration of the analog value of the YL-83 module must be done in the workplace.

B. Results:

First of all, achieve the interconnection of the different components, the different sensors have to be are calibrated. With the pH sensor, the correct calibration is checked with the pH 10.00 solution.

After checking the pH sensor calibration, the levels present in the solution are observed, this is observed with the Arduino interface. To verify the correct functioning of the implemented system, a hydroponic crop of peppers (Capsicum annuum) was implemented. The proper growth of the vegetable crop was supported with the appropriate pH level in the nutrient solution, this result was given by the pH sensor, which varies according to the type of harvest. Up next the measure of acidity or alkalinity in the crop is shown for the vegetable chosen for the validation of the proposal. In Table 1, pH levels for a pepper crop, it is indicated that the levels suitable for the pepper range from 6.5 to 7.5, the pH sensor is in charge of controlling the adequate levels in the nutrient solution based on the programming code of the microcontroller and calibration of the pH circuit.

Table 1. PH levels for a peppers crop

Type of harvest	PH Level							
Pepper (Capsicum annum)	5	5,5	6	6,5	7	7,5	8	8,5

The showed results are based on a survey with subjective questions made to experts in the area of agriculture, who compared the products obtained using the automated hydroponic system, in contrast, to products obtained by conventional sowing, with a similar period of time to develop.

4 Analysis

Due to the correct measurement of the pH sensor and humidity sensor that activate valves for automatic irrigation, we proceed to demonstrate the growth result of the pepper crop which takes about three months of development and flowering in the structure of the hydroponic system. It must be taken into consideration the rapid growth of hydroponic harvest compared to conventional crops [17, 18], due to the high proportion of air that it has, this becomes a saving of light and products, to be less time in growth than in land, with hydroponics, a larger size is reached in the same time of cultivation than on land.

Analyzing the air issues, its greater presence favors a development of the roots much higher than on land, which produces a higher production than on land, or at least, it is usually easier to get more production in hydroponics [19]

The flavors tend to be lighter, although the washing of roots in hydroponics will be much easier than on land, and, depending on the hydroponic system, it is usually also cleaner than on land. Both, at the time of the crop and at the time of throwing the surplus of the crop, in hydroponics it will be much easier to recycle the harvest products (Fig. 3).

Fig. 3. Flowering in two months of the pepper cultivation

5 Conclusions

With the construction of the structure of the hydroponic system it was possible to apply technological solutions in the field of agriculture, through a prototype applied to hydroponics with intelligent and independent automated irrigation, based on the results provided by the sensors to the microcontroller.

Not all plant crops nourish with the same pH value for their proper growth and flowering in the nutrient solution, so using the pH sensor we can apply the appropriate level of pH required by the crop in order to be able to carry out an agricultural precision. The main function is through the use of free hardware components [20] from different manufacturers that were connected together, in order to convert the analog values present in the nutritive solution into digital values sent to the microcontroller, which in turn executes the determined actions.

For the standard avr-libc programming of the source code of the Atmel micro-controller of the arduino mega, free programming software was used, such as the IDE Arduino.

It must be taken into consideration that the system has some limitations, such as the fact that a stable and continuous energy supply is required, which suggests the implementation of voltage surge protectors, against voltage variations and energy backup systems due to the sensitivity of the elements, in addition that the system is developed for a specific vegetable, as a result of the different controls implemented.

At the same time, the developed work can be improved through a macro project that seeks to implement an orchard with a greater number of vegetable varieties, with the respective expansion of sensors, research and new configuration of controllers, or on the contrary, micro projects to encourage the population to have homemade hydroponic crops as an alternative source of food for nutrition.

Acknowledgements. This work was made thanks to the contribution of Mr. Engineer Daniel Mendoza.

References

1. Diver, S.: Aquaponics-Integration of Hydroponics with Aquaculture, pp. 2–28. Attra, Melbourne (2010)
2. Jones, J.B.: Hydroponics: its history and use in plant nutrition studies. J. Plant Nutr. **5**, 1003–1030 (1982)
3. Beltrano, J., Gimenez, D.O.: Cultivo en hidroponía. D - Editorial de la Universidad Nacional de La Plata (2015)
4. Toda, T., Koyama, H., Hara, T.: A simple hydroponic culture method for the development of a highly viable root system in *Arabidopsis thaliana*. Biosci. Biotechnol. Biochem. **63**, 210–212 (1999)
5. López-Morales, V., López-Ortega, O., Ramos-Fernández, J., Muñoz, L.B.: JAPIEST: an integral intelligent system for the diagnosis and control of tomatoes diseases and pests in hydroponic greenhouses. Expert Syst. Appl. **35**, 1506–1512 (2008)
6. Domingues, D.S., Takahashi, H.W., Camara, C.A.P., Nixdorf, S.L.: Automated system developed to control pH and concentration of nutrient solution evaluated in hydroponic lettuce production. Comput. Electron. Agric. **84**, 53–61 (2012)
7. Ibayashi, H., et al.: A reliable wireless control system for tomato hydroponics. Sensors **16**, 644 (2016)
8. Saaid, M.F., Sanuddin, A., Ali, M., Yassin, M.S.A.I.M.: Automated pH controller system for hydroponic cultivation. In: 2015 IEEE Symposium on Computer Applications & Industrial Electronics (ISCAIE), pp. 186–190. IEEE (2015)
9. Ferentinos, K.P., Albright, L.D., Selman, B.: Neural network-based detection of mechanical, sensor and biological faults in deep-trough hydroponics. Comput. Electron. Agric. **40**, 65–85 (2003)
10. Guo, X., van Iersel, M.W., Chen, J., Brackett, R.E., Beuchat, L.R.: Evidence of association of salmonellae with tomato plants grown hydroponically in inoculated nutrient solution. Appl. Environ. Microbiol. **68**, 3639–3643 (2002)
11. Savvas, D., et al.: Interactions between salinity and irrigation frequency in greenhouse pepper grown in closed-cycle hydroponic systems. Agric. Water Manag. **91**, 102–111 (2007)
12. Tuberosa, R., Sanguineti, M.C., Landi, P., Michela Giuliani, M., Salvi, S., Conti, S.: Identification of QTLs for root characteristics in maize grown in hydroponics and analysis of their overlap with QTLs for grain yield in the field at two water regimes. Plant Mol. Biol. **48**, 697–712 (2002)
13. Wong, J.: United States Patent
14. Powering the arduino with a 5 V power supply. https://forum.arduino.cc/index.php?topic=271158.0
15. Banzi: Getting Started with Arduino and Genuino products. (2017)
16. Strohbach, M., Gellersen, H.-W., Kortuem, G., Kray, C.: Cooperative Artefacts: Assessing Real World Situations with Embedded Technology. Presented at the (2004)
17. Llamas, L.: Medir la humedad del suelo con Arduino y sensor FC-28. https://www.luisllamas.es/arduino-humedad-suelo-fc-28/
18. Inga, F.: Diferencias entre cultivo hidroponico y cultivo en tierra. https://www.paisagrowseeds.com/diferencias-entre-cultivo-hidroponico-y-cultivo-en-tierra/

19. Jafarnia, S., Khosrowshahi, S., Hatamzadeh, A., Ali, T.: Effect of substrate and variety on some important quality and quantity characteristics of strawberry production in vertical hydroponics system. Adv. Environ. Biol. **3**, 360–363 (2010)
20. Harnett, C.: Open source hardware for instrumentation and measurement. IEEE Instrum. Meas. Mag. **14**, 34–38 (2011)

IoT Applications in Agriculture: A Systematic Literature Review

Raquel Gómez-Chabla[1(✉)], Karina Real-Avilés[1], César Morán[2], Paola Grijalva[1], and Tanya Recalde[3]

[1] Escuela de Ingeniería en Computación e Informática, Guayaquil, Ecuador
{rgomez, kreal, pgrijalva}@uagraria.edu.ec
[2] Carrera de Ingeniería Agronómica, Facultad de Ciencias Agrarias, Universidad Agraria del Ecuador, Av. 25 de Julio y Pio Jaramillo, P.O. BOX 09-04-100, Guayaquil, Ecuador
cmoran@uagraria.edu.ec
[3] Carrera de Medicina, Facultad de Ciencias Médicas, Universidad de Guayaquil, Salvador Allende entre, 1er Callejón 5 No, P. O, BOX 09-06-13, Guayaquil, Ecuador
tanya.recaldec@ug.edu.ec

Abstract. The digital breach between agricultural producers and IoT technologies has reduced in the last years. In the future, these technologies will allow improving productivity through the sustainable cultivation of food, as well as to take care of the environment thanks to the efficient use of water and the optimization of inputs and treatments. IoT technologies allow developing systems that support different agricultural processes. Some of these systems are remote monitoring systems, decision support tools, automated irrigation systems, frost protection systems, and fertilization systems, among others. Considering the aforementioned facts, it is necessary to provide farmers and researchers with a clear perspective of IoT applications in agriculture. In this sense, this work presents a systematic literature review of IoT-based tools and applications for agriculture. The objective of this paper is to offer an overview of the IoT applications in agriculture through topics such IoT-based software applications for agriculture available in the market, IoT-based devices used in the agriculture, as well as the benefits provided by this kind of technologies.

Keywords: IoT · Agriculture · Software · Sensors · Cloud computing

1 Introduction

The advances in science and technology and the high qualification of human capital have allowed a sustainable growth of the world economy [1–3]. This fact has resulted in the emergence of the smart farming approach which allows farmers to remotely monitor the crop field by means of sensors as well as to have automatic irrigation systems [4]. There are computer applications based on sensors that allow obtaining more accurate information about the crop, soil, and climate than those obtained by means of traditional methods. This feature helps to improve the quality of the products,

R. Valencia-García et al. (Eds.): CITAMA 2019, AISC 901, pp. 68–76, 2019.
https://doi.org/10.1007/978-3-030-10728-4_8

processes and raw materials used in this process. Because of these facts, IoT-based smart agriculture is more efficient than traditional approaches [5]. Furthermore, the IoT-based smart agriculture applications could boost organic agricultural agriculture and family farming [6].

The digital breach between agricultural producers and IoT technologies [7] has reduced. In the future, these technologies will allow improving productivity through the sustainable cultivation of food, as well as to take care of the environment thanks to the efficient use of water and the optimization of inputs and treatments [8]. Smart agriculture includes activities such as remote monitoring, decision support tools, automated irrigation systems, frost protection, fertilization, among others. These activities are supported by IoT technologies such as hardware [9, 10], intelligent software [11], integration platforms [12], monitoring processes [13, 14], operating systems, and cloud computing [15]. The Cloud of Things [16], which is the integration of IoT and cloud computing, can help achieve the objectives of the IoT and the Internet [17]. Furthermore, the IoT must help society to have information transparency [18, 19].

This work presents a systematic literature review of IoT-based tools and applications for agriculture. The objective of this paper is to offer an overview of these areas through topics such as IoT-based software applications for agriculture available in the market, IoT-based devices used in the agriculture, as well as the benefits provided by these technologies.

2 IoT Applications in Agriculture

This section describes the literature review process followed in this work which is composed of three steps: question formulation, search strategy, and studies selection. This review analyzes research efforts and projects related to the application of IoT technologies to the field of agriculture aiming to provide guidelines for the use of these technologies in urban agriculture, precision agriculture and industrial agriculture.

2.1 Research Questions

The research questions that guided this work, as well as their motivation, are presented in Table 1.

Table 1. Research questions

RQ	Question	Motivation
RQ1	What are the specific areas of agriculture where IoT technologies have been applied?	To detect the main areas of agriculture where IoT technologies are used
RQ2	What IoT-based devices are used in agriculture?	To identify the main IoT-based devices used in agriculture
RQ3	What are the IoT-based software applications for agriculture?	To identify IoT-based software applications used in agriculture
RQ4	What are the benefits of IoT in agriculture?	To identify the main benefits of IoT in agriculture

2.2 Search Strategy

We identified the digital libraries wherein the search for primary studies was performed: IEEE CS, Elsevier, ACM Digital Library, and ScienceDirect. Then, we identified a set of keywords related to our research topic: environment, apps, devices, IoT, Internet of Things, Intelligent agriculture, cloud in the agriculture, and analytics data in the agriculture. Finally, the search string was built by combining the aforementioned keywords with the connectors "AND" and "OR". Thus, the search chain that we use is presented in Table 2.

Table 2. Keywords and data sources used in this literature review

Source	Search string	Context
IEEE CS	(("environment" OR "Apps" OR "Devices") AND ("IoT" OR "Internet of Things" OR "Intelligent Agriculture" OR "Cloud in the Agriculture") OR ("Analytics Data in the Agriculture"))	Agriculture
Elsevier		
ACM digital library		
ScienceDirect		

2.3 Studies Selection

The studies selection process consisted of executing the search for studies in the digital libraries considering the search string as well as some inclusion and exclusion criteria. Specifically, this process consisted of four phases which are described below. Also, Table 2 presents a summary of the studies selection phases performed in this work.

Table 3. Studies selection

Phase	IEEE CS	Elsevier	ACM	ScienceDirect	Total
1	1009	800	500	205	2514
2	950	700	420	78	2148
3	99	75	300	35	509
4	26	13	17	4	60

1. This phase consisted of executing the search for studies by using the search string presented in Table 2. Furthermore, we apply the following exclusion criteria: (1) papers written in English, and (2) papers published in the period of 2011–2018. The result of this phase was a set of 2514 studies.
2. Secondly, we discarded those papers whose title was not directly related to the IoT technology and the agricultural sector. The result of this phase was a set of 2148 studies.
3. Thirdly, the set of papers obtained in the previous phase was reduced by discarding those papers whose abstract was not directly related to the IoT technology and the agricultural sector. The result of this phase was a set of 509 studies.

4. Finally, in cases where we were uncertain about the relevance of the paper, the full-paper was downloaded and sections such as introduction and conclusions were analyzed. The objective of this phase was to determine which piece of literature provide important contributions to the agricultural sector. The result of this phase was a set of 60 studies.

3 Information Extraction

This section provides a general perspective of all studies selected aiming to answer the research questions established at the beginning of this work.

3.1 IoT Applications in Agriculture

The main applications of IoT technologies in agriculture are found in precision agriculture [20, 21] whose architecture includes IoT techniques for urban agriculture and precision agronomy in smart cities. Commonly, smart cities are based on networks defined by software (SDN) and cyber-physical systems [22]. Other applications of the IoT are the agricultural drones [23] which are relatively cheap drones with advanced sensors that give farmers new ways to increase yields and reduce crop damage, among other things. Another area of IoT application is the intelligent greenhouses [24, 25] which includes hydroponic and small-scale aquaponic systems [26–28]. Intelligent greenhouses are increasingly common in urban areas because they allow monitoring several parameters of nutrient solutions [29], as well as to improve the growth, yield, and quality of plants. These improvements contribute significantly to the achievement of smart cities with infrastructures that allow automating, optimizing and improving urban agriculture and precision agronomy. Another area in which IoT technologies are applied is the vertical agriculture [30], which allows controlling soil moisture and water content by means of computers or mobile devices such as tablets and smartphones. Finally, there are applications that combine IoT technologies with Artificial Intelligence such as Malthouse [31], which is an Artificial Intelligence system that allows prescribing configurations and schedules in precision farming and food manufacturing areas.

3.2 IoT-Based Devices Used in Agriculture

IoT-based devices have been adopted by many industries and markets around the world. One of these industries is agriculture, which benefits from IoT technologies in a myriad of ways. For example, LoRa is a widely used network radio in Indiana because of its advantages such as long range, low power consumption, and its low-cost investment [32, 33]. Another example of the use of IoT-based devices is the use of cameras [34, 35] to verify the quality of food [36]. On the other hand, there are approaches that combine cloud computing with wireless sensor networks to offer a service (AaaS, Agriculture-as-a-Service) that allows managing agricultural information through Big Data technologies [37]. For example, the Phytec company offers the

PlanIoT platform that is capable of automatically detecting plant status changes, analyzing information, and generating recommendations.

3.3 IoT-Based Software Applications Used in Agriculture

IoT-based technologies have been successfully adopted in different contexts. Due to this fact, several companies are investing in IoT-based software development for agriculture. Nowadays, there are several software products available in the market focused on supporting different agricultural processes. For instance, AG-IoT [23] is an unmanned aerial vehicle that locates and assists IoT-based devices available on the ground to form groups for the transmission of data. On the other hand, Agro 4.0 [38] implements high-performance computational methods, a sensors network, connectivity between mobile devices, cloud computing, and analytical methods to process large volumes of data and provide decision support systems. Agro-Tech [39] records, stores, and updates the data obtained from various sensors available in a specific area of the crop. Also, this software allows farmers to access this information aiming to monitor their crop. Malthouse [31] is an Artificial Intelligence system that allows prescribing configurations and schedules in precision farming and food manufacturing areas.

There are agricultural monitoring systems that transmit live video to carry out such process remotely through IoT-based devices that integrate cameras and Raspberry Pi cards [35, 40]. On the other hand, Cropx is an adaptive irrigation software tool that allows farmers to increase crop yields at the lowest cost possible, as well as saving water and energy. Farmlogs is a farm management software that allows registering activities related to the conservation of crop through images. Finally, MbeguChoice is an application that allows farmers to select the best drought tolerant seed suppliers.

3.4 Benefits of IoT in Agriculture

The main benefits of IoT in agriculture identified in this literature review are briefly described below.

- Community agriculture in urban and rural areas taking advantage of hardware and software resources and large amounts of data.
- Logistic and qualitative traceability of food production that allows reducing costs and the waste of inputs through the use of real-time data for decision making.
- Generation of business models [31] in the agricultural context that allow establishing a direct relationship with the consumer.
- Crop monitoring that allows reducing costs as well as the theft of machinery.
- Automatic irrigation systems [41] that work according to temperature, humidity, and soil moisture values that are obtained through sensors.
- Automatic collection of environmental parameters through sensor networks for further processing and analysis.
- Decision support systems that analyze large amounts of data to improve operational efficiency and productivity [42, 43].

4 Discussion

Table 3 shows a summary of the studies selected in this work. As can be seen, 21 articles address the application of IoT technologies in agriculture. 12 of the selected studies address the use of IoT-based devices in agriculture [44]. Some of the most common devices used in this context are optical sensors to measure soil properties, photodiodes and photodetectors to determine the soil, organic matter, and soil moisture, moisture sensors to measure the amount of water in the soil, and geo-positioning devices to determine latitude, length, and altitude. It is worth mentioning that geo-positioning devices are an important element in precision agriculture [45]. Other IoT-based devices widely used in the agricultural context are cameras and wireless networks [32], which are used for the development of platforms for obtaining and analyzing information. On the other hand, some of the most outstanding software tools identified in this work are a support system for renewable energy projects, systems focused on resources and ecological techniques, decision support systems for precision agriculture [46, 47], systems based on wireless sensor networks [39], and intelligent systems for vertical agriculture [12] (Table 4).

Table 4. Summary of selected studies

Aim	Works
IoT applications in agriculture	21
IoT-based devices used in agriculture	12
IoT-based software applications used in agriculture	8
Advantages	11
Others	8
Total	60

5 Conclusions

IoT technologies allow, among other things, to obtain information on climate, humidity, temperature, soil fertility in order to efficiently carry out remote monitoring of crops. Thanks to these technologies, farmers can know the status of their crop at any time and from any place. On the other hand, wireless sensor networks allow controlling the conditions of the farm, as well as automate different processes. For example, some of the studies analyzed in this work use wireless cameras to know the status of the crop in real time. Other studies have employed drones to support the tasks of precision agriculture, as well as smartphones to keep farmers informed about the current conditions of their cultivation. Some of the most outstanding technologies that are combined with IoT to develop agricultural solutions are wireless sensor networks, cloud computing, middleware systems, and mobile applications [48, 49]. IoT technologies are already an essential part of solving different problems in the agricultural context. For this reason, in this work, we performed a literature review aiming to identify the main IoT applications in agriculture, IoT-based software and devices used in agriculture, as well as the benefits provided by this kind of technologies. Although the topics addressed in this work are very important for all people involved in agriculture, it is important to mention the need for analyzing more research works carried out around the resolution of environmental problems aiming to achieve a sustainable cultivation of food.

References

1. Gómez-Chabla, R., Aguirre-Munizaga, M., Samaniego-Cobo, T., Choez, J., Vera-Lucio, N.: A reference framework for empowering the creation of projects with arduino in the ecuadorian universities. In: Valencia-García, R., Lagos-Ortiz, K., Alcaraz-Mármol, G., Del Cioppo, J., Vera-Lucio, N., Bucaram-Leverone, M. (eds.) Communications in Computer and Information Science, pp. 239–251 (2017)
2. Bouaziz, M., Rachedi, A.: A survey on mobility management protocols in Wireless Sensor Networks based on 6LoWPAN technology. Comput. Commun. **74**, 3–15 (2016)
3. van der Ploeg, J.D.: Peasant-driven agricultural growth and food sovereignty. J. Peasant Stud. **41**, 999–1030 (2014)
4. Goumopoulos, C., O'Flynn, B., Kameas, A.: Automated zone-specific irrigation with wireless sensor/actuator network and adaptable decision support. Comput. Electron. Agric. **105**, 20–33 (2014)
5. Bertino, E., Choo, K.-K.R., Georgakopolous, D., Nepal, S.: Internet of things (IoT). ACM Trans. Internet Technol. **16**, 1–7 (2016)
6. Berte, D.-R.: Defining the IoT. Proc. Int. Conf. Bus. Excell. **12**, 118–128 (2018)
7. Sugawara, E., Nikaido, H.: Properties of AdeABC and AdeIJK efflux systems of Acinetobacter baumannii compared with those of the AcrAB-TolC system of Escherichia coli. Antimicrob. Agents Chemother. **58**, 7250–7257 (2014)
8. Phupattanasin, P., Tong, S.-R.: Applying information-centric networking in today's agriculture. APCBEE Procedia **8**, 184–188 (2014)
9. Akhtar, P., Khan, Z., Tarba, S., Jayawickrama, U.: The Internet of Things, dynamic data and information processing capabilities, and operational agility. Technol. Forecast. Soc. Change (2017)
10. Hlaing, W., Thepphaeng, S., Nontaboot, V., Tangsunantham, N., Sangsuwan, T., Pira, C.: Implementation of WiFi-based single phase smart meter for Internet of Things (IoT). In: 2017 International Electrical Engineering Congress (iEECON), pp. 1–4. IEEE (2017)
11. Venkatesan, R., Tamilvanan, A.: A sustainable agricultural system using IoT. In: 2017 International Conference on Communication and Signal Processing (ICCSP), pp. 0763–0767. IEEE (2017)
12. González, D.R.: Arquitectura y Gestión de la IoT. Rev. Telem@tica, 12 (2013)
13. Mohd Kassim, M.R., Mat, I., Harun, A.N.: Wireless sensor network in precision agriculture application. In: 2014 International Conference on Computer, Information and Telecommunication Systems (CITS), pp. 1–5. IEEE (2014)
14. Li, S.: Application of the internet of things technology in precision agriculture IRrigation systems. In: 2012 International Conference on Computer Science and Service System, pp. 1009–1013. IEEE (2012)
15. Aqeel-ur-Rehman: Smart Agriculture. In: 2017 International Conference on Innovations in Electrical Engineering and Computational Technologies (ICIEECT), pp. 1–1. IEEE (2017)

16. Carrasquilla-Batista, A., Chacon-Rodriguez, A., Solorzano-Quintana, M., Guerrero-Barrantes, M.: IoT applications: on the path of Costa Rica's commitment to becoming carbon-neutral. In: 2017 International Conference on Internet of Things for the Global Community (IoTGC), pp. 1–6. IEEE (2017)
17. Perera, C., Zaslavsky, A., Christen, P., Georgakopoulos, D.: Context aware computing for the internet of things: a survey. Cogn. Neuropsychol. **29**, 349–353 (2013)
18. Chen, H.-C., Chang, C.-H., Leu, F.-Y.: Implement of agent with role-based hierarchy access control for secure grouping IoTs. In: 2017 14th IEEE Annual Consumer Communications & Networking Conference (CCNC), pp. 120–125. IEEE (2017)
19. Sosa, E., Godoy, D.: Internet del Futuro. Desafíos y perspectivas. Rev. Cienc. y Tecnol. **1**, 40–46 (2013)
20. Mekala, M.S., Viswanathan, P.: A survey: smart agriculture IoT with cloud computing. In: 2017 International conference on Microelectronic Devices, Circuits and Systems (ICMDCS), pp. 1–7. IEEE (2017)
21. Rajeswari, S., Suthendran, K., Rajakumar, K.: A smart agricultural model by integrating IoT, mobile and cloud-based big data analytics. In: 2017 International Conference on Intelligent Computing and Control (I2C2), pp. 1–5. IEEE, Coimbatore (2017)
22. Ordonez-Garcia, A., Siller, M., Begovich, O.: IoT architecture for urban agronomy and precision applications. In: 2017 IEEE International Autumn Meeting on Power, Electronics and Computing (ROPEC), pp. 1–4. IEEE (2017)
23. Uddin, M.A., Mansour, A., Le Jeune, D., Aggoune, E.H.M.: Agriculture internet of things: AG-IoT. In: 2017 27th International Telecommunication Networks and Applications Conference (ITNAC), pp. 1–6. IEEE (2017)
24. Wortman, S.E.: Crop physiological response to nutrient solution electrical conductivity and pH in an ebb-and-flow hydroponic system. Sci. Hortic. (Amsterdam) **194**, 34–42 (2015)
25. Yunseop, K., Evans, R.G., Iversen, W.M.: Remote sensing and control of an irrigation system using a distributed wireless sensor network. IEEE Trans. Instrum. Meas. **57**, 1379–1387 (2008)
26. Atmadja, W., Liawatimena, S., Lukas, J., Nata, E.P.L., Alexander, I.: Hydroponic system design with real time OS based on ARM Cortex-M microcontroller. In: IOP Conference Series: Earth and Environmental Science (2018)
27. Al-Karaki, G.N., Al-Hashimi, M.: Green fodder production and water use efficiency of some forage crops under hydroponic conditions. ISRN Agron. **2012**, 1–5 (2012)
28. Montoya, A.P., Obando, F.A., Morales, J.G., Vargas, G.: Automatic aeroponic irrigation system based on Arduino's platform. In: Journal of Physics: Conference Series (2017)
29. Barbosa, G.L., et al.: Comparison of land, water, and energy requirements of lettuce grown using hydroponic vs. Conventional agricultural methods. Int. J. Environ. Res. Public Health **12**(6), 6879–6891(2015)
30. Bin Ismail, M.I.H., Thamrin, N.M.: IoT implementation for indoor vertical farming watering system. In: 2017 International Conference on Electrical, Electronics and System Engineering (ICEESE), pp. 89–94. IEEE (2017)
31. Dolci, R.: IoT solutions for precision farming and food manufacturing: artificial intelligence applications in digital food. In: 2017 IEEE 41st Annual Computer Software and Applications Conference (COMPSAC), pp. 384–385. IEEE (2017)
32. Jin, J., Ma, Y., Zhang, Y., Huang, Q.: Design and implementation of an agricultural IoT based on LoRa. MATEC Web Conf. **189**, 04011 (2018)
33. Webb, J., Hume, D.: Campus IoT collaboration and governance using the NIST cybersecurity framework. In: Living in the Internet of Things: Cybersecurity of the IoT – 2018, p. 26 (7 pp.). Institution of Engineering and Technology (2018)

34. Krishna, K.L., Silver, O., Malende, W.F., Anuradha, K.: Internet of Things application for implementation of smart agriculture system. In: 2017 International Conference on I-SMAC (IoT in Social, Mobile, Analytics and Cloud) (I-SMAC), pp. 54–59. IEEE (2017)
35. Anvekar, R.G., Banakar, R.M., Bhat, R.R.: Design alternatives for end user communication in IoT based system model. In: 2017 IEEE Technological Innovations in ICT for Agriculture and Rural Development (TIAR), pp. 121–125. IEEE (2017)
36. Mohanraj, I., Ashokumar, K., Naren, J.: Field monitoring and automation using IOT in agriculture domain. Procedia Comput. Sci. **93**, 931–939 (2016)
37. Gill, S.S., Chana, I., Buyya, R.: IoT based agriculture as a cloud and big data service. J. Organ. End User Comput. **29**, 1–23 (2017)
38. Fonseca, S.M., Massruhá, S., Angelica De Andrade Leite, M.: Agro 4.0 – Rumo À Agricultura Digital, pp. 28–35 (2016)
39. Pandithurai, O., Aishwarya, S., Aparna, B., Kavitha, K.: Agro-tech: a digital model for monitoring soil and crops using internet of things (IOT). In: 2017 Third International Conference on Science Technology Engineering & Management (ICONSTEM), pp. 342–346. IEEE (2017)
40. Shete, R., Agrawal, S.: IoT based urban climate monitoring using raspberry Pi. Int. Conf. Commun. Signal Process. April 6–8, 2016, India IoT. 2008–2012 (2016)
41. Kaewmard, N., Saiyod, S.: Sensor data collection and irrigation control on vegetable crop using smart phone and wireless sensor networks for smart farm. In: 2014 IEEE Conference on Wireless Sensors (ICWiSE), pp. 106–112. IEEE (2014)
42. Nandyala, C.S., Kim, H.K.: Green IoT agriculture and healthcare application (GAHA). Int. J. Smart Home **10**, 289–300 (2016)
43. Cambra, C., Sendra, S., Lloret, J., Garcia, L.: An IoT service-oriented system for agriculture monitoring. In: 2017 IEEE International Conference on Communications (ICC), pp. 1–6. IEEE (2017)
44. Ma, J., Zhou, X., Li, S., Li, Z.: Connecting agriculture to the internet of things through sensor networks. In: 2011 International Conference on Internet of Things and 4th International Conference on Cyber, Physical and Social Computing, pp. 184–187. IEEE (2011)
45. Stewart, J., Stewart, R., Kennedy, S.: Internet of things — propagation modelling for precision agriculture applications. In: 2017 Wireless Telecommunications Symposium (WTS), pp. 1–8. IEEE (2017)
46. Kiani, F., Seyyedabbasi, A.: Wireless sensor network and internet of things in precision agriculture. Int. J. Adv. Comput. Sci. Appl. **9**(8), 220–226 (2018)
47. Kapoor, A., Bhat, S.I., Shidnal, S., Mehra, A.: Implementation of IoT (Internet of Things) and image processing in smart agriculture. In: 2016 International Conference on Computation System and Information Technology for Sustainable Solutions (CSITSS), pp. 21–26. IEEE (2016)
48. Elijah, O., Rahman, T.A., Orikumhi, I., Leow, C.Y., Hindia, M.N.: An overview of internet of things (IoT) and data analytics in agriculture: benefits and challenges. IEEE Internet Things J. 1–1 (2018)
49. Sales, N., Remedios, O., Arsenio, A.: Wireless sensor and actuator system for smart irrigation on the cloud. In: IEEE World Forum on Internet of Things, WF-IoT 2015 - Proceedings, pp. 693–698. IEEE (2015)

Image Processing

Analysis of Computer Vision Algorithms to Determine the Quality of Fermented Cocoa (Theobroma Cacao): Systematic Literature Review

Karen Mite-Baidal(✉), Evelyn Solís-Avilés, Tayron Martínez-Carriel, Augusto Marcillo-Plaza, Elicia Cruz-Ibarra, and Wilmer Baque-Bustamante

Faculty of Agricultural Sciences, Computer Science Department, Agrarian University of Ecuador, Av. 25 de Julio y Pio Jaramillo, P.O. BOX 09-04-100, Guayaquil, Ecuador
{kmite, esolis, tmartinez, jmarcillo, ecruz, wbaque}@uagraria.edu.ec

Abstract. Computer vision techniques have been used for the automation of processes in the agricultural sector due to the benefits obtained such as effectiveness and quality. A clear example is the analysis of cocoa beans quality. The increasing interest of computer vision in this area calls for a clear, systematic overview. In this sense, we present a systematic literature review (SLR) of computer vision algorithms to determine the quality of fermented cocoa in a six-year period: from 2013–2018. The aim of this review is to identify the techniques of computer vision algorithms used to assess fermentation index of cocoa beans for quality control, as well, the main physical and chemical characteristics of the cocoa beans identified through the computer vision algorithms. The results show that the PLS (Partial Least-Squares) algorithm is the most used for the classification of images in a statistical approach. Also, color is the physical parameter that is commonly identified through artificial vision algorithms. Meanwhile, Fat and pH are the chemical parameters most identified by FT-NIR (Fourier transform near-infrared) technology in conjunction with the chemometric technique.

Keywords: Computer vision · Computer vision algorithms · Image recognition

1 Introduction

Quality control and food safety are critical concerns of food manufacturers, governments and consumers because there have been many outbreaks of diseases in recent years. Because of the processing in cacao bean manufacturing, including heat treatment and the removal of excess moisture, a foodborne illness from chocolate products is comparatively less likely. However, the cocoa, candy, and chocolate industry face the challenge of ensuring that raw materials, including cacao beans, are of high quality and safe. Cacao beans are spontaneously fermented seeds, and as such are subject to a high

R. Valencia-García et al. (Eds.): CITAMA 2019, AISC 901, pp. 79–87, 2019.
https://doi.org/10.1007/978-3-030-10728-4_9

level of variability depending on growing conditions, genetics, postharvest fermentation and drying of the cocoa beans prior to shipment or handling. This variability can have a large impact on the finished chocolate product, therefore, it is important to be able to determine if cacao beans are properly fermented, of high quality, and lacking defects [1].

Computer vision techniques have been used for the automation of processes in the agricultural sector due to the benefits obtained such as effectiveness and quality [2, 3]. Recently several authors have focused on computer vision algorithms to determine the quality of fermented cocoa. Therefore, it is important to identify what computer vision algorithms have been already studied for the image processing in the fermentation of cocoa and what are the physical and chemical characteristics of the cocoa beans identified. To address these questions, we presented a Systematic Literature Review (SLR) to identify relevant papers related to the computer vision in agriculture. The systematic literature review presented was performed considering published papers from 2013 to 2018. This information could help other researchers in identifying possible research areas for future research as well as farmers to know automated processes that are being used.

The rest of this paper is organized as follows: Sect. 2 presents the analysis and review planning which is divided into four parts: research questions, search sources, search strategy and exclusion criteria. Section 3 presents the results of the systematic literature review. Section 4 presents a discussion of the obtained results. Finally, our conclusions are presented in Sect. 5.

2 Analysis and Review Planning

The main goal of this Systematic Literature Review is to identify the techniques or computer vision algorithms used to assess fermentation index of cocoa beans for quality control. Also, to identify which image processing technique within computer vision is the most applied today in this area.

2.1 Research Questions

This section presents the three research questions that guided us throughout the research and helped us meet the goals of the Systematic literature review.

Q1: What are the image processing levels in a computer vision model?

Q2: What are the physical and chemical characteristics of the cocoa beans that identify the computer vision algorithms?

Q3: What are the computer vision algorithms that are used for the image processing in the fermentation of cocoa?

2.2 Search Sources

Table 1 shows the search sources from which the systematic literature review was conducted. The first column represents the digital libraries wherein the search for primary studies would be carried out. The second column indicates the type of publication. The third column shows the criteria about will conduct the search. The last one represents the language.

Table 1. Search sources

Digital libraries	Publication type	Search based on	Language
IEEE Xplore	Book	Keywords	English
Science direct	Journals		
Springer link	Conferences		
Wiley	Journals		
Google scholar			
ERIC institute of education of science			
Elsevier			
ACM DL digital library			
Publication year	2013–2018		

2.3 Search Strategy

A search strategy was conducted to respond to the questions presented in Sect. 2.1. The search strings were built by combining the keywords with the connectors "AND" and "OR". Thus, the search chain that we use is the following:

(Determination of Cocoa Bean Theobroma cacao Quality with Image Processing) AND (tool to determine cocoa fermentation levels) OR (Computer vision for external quality of Theobroma cacao) AND/OR (Estimating cocoa bean parameters by computer vision) OR (Image Processing).

We have extracted studies from scientific databases published between 2013 and 2018.

2.4 Exclusion Criteria

The set of results obtained was reduced by applying the following exclusion criteria:

- Papers that despite having the phrase artificial vision in their content, they did not provide an important contribution to the subject of the review.
- Papers that analyze the quality of cocoa through manual or traditional methods.
- Papers that analyze computer vision algorithms for fruits or any food.

3 Review Execution

This section presents the results of the systematic literature review and it is divided into two subsections, depending on the research question that is addressed. The first subsection discusses the image processing levels in a computer vision model. Finally, the second subsection presents the main physical and chemical characteristics of the cocoa bean that identify the computer vision algorithms, as well as, the computer vision algorithms that are used for the image processing in the fermentation of cocoa.

3.1 Q1: What Are the Image Processing Levels in a Computer Vision Model?

The objective of this research question was to identify the main levels in a computer vision model.

Computer Vision. The techniques, algorithms, and applications of computer vision or artificial vision allow acquiring, processing, analyzing and understanding the images in order to produce numerical or symbolic information used for decision making both in real time and in later analyzes [4]. In addition, they are used in positions where environmental and safety conditions would not allow human presence.

The traditional computer vision system is usually based on RGB color cameras that mimic the vision of the human eye by capturing images using three filters centered on the red, green and blue (RGB) wavelengths.

Hyperspectral computer vision systems are used to acquire images with high spatial and spectral resolutions. A typical hyperspectral imaging (HSI) system consists of the following components: a light source (illumination), a wavelength dispersion device (spectrograph), an area detector (camera), the translation stage and a computer. This type of system is applied for some fundamental investigations such as quality assessment of agricultural products [5].

Multispectral computer vision system is a form of imaging that involves capturing two or more different wavelength monochromatic images in the spectrum. The multispectral computer vision system, such as hyperspectral ones, it is applied to evaluate the safety and quality of foods such as the study for automatic classification of bicolored apples by means of multispectral artificial vision and it is also used for fast online applications [6].

Image Processing Levels. Nowadays, there are tools and applications that include analysis algorithms and techniques for image processing or recognition. These techniques include terrestrial or aerial remote sensing that by capturing images using thermal, hyperspectral and photometric cameras (RGB-D), those are processed and analyzed to solve problems in different areas of agriculture [7].

In the image processing and analysis, there are development and applications of computer vision for the external quality inspection of fruits and vegetables. Zhang et al. [8] establishes three levels: (1) low-level processing, which involves the image acquisition and the image preprocessing; (2) intermediate level (image processing and analysis) that involves the image segmentation, extraction of characteristics, representation, and description; and (3) high level (key step of image analysis), which implies recognition, interpretation, and classification. Next, Fig. 1 shows the levels of image processing.

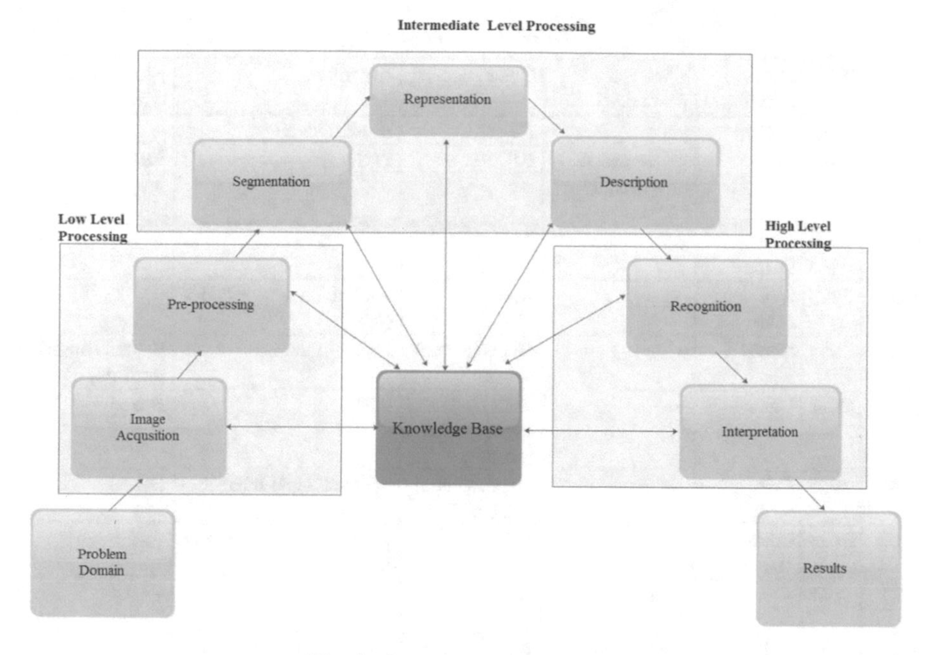

Fig. 1. Image processing levels

3.2 Q2: What Are the Physical and Chemical Characteristics of the Cocoa Beans that Identify the Computer Vision Algorithms? Q3: What Are the Computer Vision Algorithms That Are Used for the Image Processing in the Fermentation of Cocoa?

To respond to these research questions, Table 2 presents the computer vision algorithms that are used for the image processing in the fermentation of cocoa, as well as, the main physical and chemical characteristics of the cocoa beans identified through the computer vision algorithms.

Table 2. Physical, chemical parameters identified by computer vision algorithms according to image segmentation by similarity

Paper	Year	Physical parameters	Chemical parameters	Color model	Algorithm	Fermentation type
[9]	2016	Color	Sugar free amino acids	RGB.	Artificial neural network (ANN)	Boxed (6 days 40 °C) -sun-dried
[10]	2015	–	Total fat	FT-NIR	Support vector machine regression (SVMR)	–
[11]	2014	–	NH_3	NIR	Modified Partial least-squares regression (MPLS)	Boxed (6 days +20 °C) -sun-dried

(*continued*)

Table 2. (*continued*)

Paper	Year	Physical parameters	Chemical parameters	Color model	Algorithm	Fermentation type
[12]	2015	Color	pH	FT-NIR	Si-BPANNR	–
[13]	2016	Color	pH and total polyphenol content	FT-NIR	PLS	Fermentation by heap method
[14]	2018	Size Shape	–	RGB	MELS-SVM	–
[15]	2013	Size and color	–	RGB	Artificial neural network (ANN)	–
[16]	2015	Size	–	Hyperspectral images	Fuzzy logic	Boxed. sun-dried for 5 days
[17]	2017	Color	–	RGB hyperspectral images	KNN	Boxed
[18]	2015	Color	Anthocyanin	Hyperspectral images	Spectral angle mapper (SAM).	Boxed
[19]	2018	Color	–	RGB	Artificial neural network (ANN)	–
[20]	2018	–	Humidity pH acidity fat	NIR	PLS	–
[21]	2016	Color	Anthocyanin	Hyperspectral images	Spectral angle mapper (SAM)	Boxed
[22]	2018	Fungi	–	FT-NIR	PLS	–
[23]	2014	–	Polyphenols	NIR	SVMR	–
[24]	2017	–	Volatile organic compounds	NIR	KNN	Fermentation by heap method
[25]	2018	Color	–	RGB	KNN	Boxed
[26]	2013	–	Xanthines (Theobromine) Polyphenols	FT-NIR	PLS	–

4 Result Analysis: Discussion

The cocoa market establishes minimum physical and chemical parameters that the grains must meet for commercialization. The classification of the cocoa in the producing countries is based on the visual estimation of the grain quality by means of a procedure known as "test of the cut" described in the Colombian Technical Standard 1252. This procedure was applied to the study of cultivated cacao in Colombia based on standardized methods, quality indicators and variety [27]. Among the physical and chemical parameters are the grain index (determines the average weight of the grains in grams), the percentage of humidity, percentage of fat, PH and the cob index (number of cobs that are needed to obtain one kg of dry cocoa). These parameters are detected through the computer vision algorithms together with the chemometric technique.

Spectral and image information obtained by computer vision systems such as RGB, hyperspectral, multispectral, near-infrared spectroscopy, Fourier Transform Near Infrared (FT-NIR) and chemometric techniques allow to extract or discover some characteristics of cocoa that are impossible or difficult to detect through human vision.

Color is the physical parameter that is commonly identified through artificial vision algorithms. Meanwhile, Fat and pH are the chemical parameters most identified by FT-NIR (Fourier transform near-infrared) technology in conjunction with the chemometric technique.

On the other hand, the PLS (Partial Least-Squares) algorithm is the most used for image classification [28]. PLS correlates the spectral variation with the variation of the concentration of components to create component prediction models. Also, the results presented in Table 2 show that the KNN algorithm [29] and the Neural Networks [30] are also used to improve the accuracy of hyperspectral image classification.

There is a research gap since the physical and chemical parameters are factors influenced by the type of cocoa, the chemical compositions of the soil, the type of fermentation and drying.

5 Conclusions and Future Work

The goal of this study was to identify the techniques or computer vision algorithms used to assess fermentation index of cocoa beans for quality control, as well, the main physical and chemical characteristics of the cocoa beans identified through the computer vision algorithms. We obtained and analyzed 18 studies from scientific databases and web sources.

There are several algorithms usually studied in the literature: Artificial Neural Network, Support Vector Machine Regression, Modified Partial least-squares regression, Partial least-squares regression, Spectral Angle Mapper, and k-nearest neighbors. Partial Least-Squares algorithm is the most used for the classification of images in a statistical approach. Most of the studies are based in identify physical parameters such as color, shape, size and Fungi. Although, color is the physical parameter that is commonly identified through artificial vision algorithms. As for the chemical parameters, we found that Fat and pH are the most identified by FT-NIR (Fourier transform near-infrared) technology in conjunction with the chemometric technique.

As future work, we plan to extend this work by including a wider set of digital libraries such as Web of science. Furthermore, we expect this systematic literature review to include more issues and proposed solutions to overcome challenges and limitations of computer vision technology. Finally, we plan to evaluate works that fulfill two main characteristics: 1) works oriented to the creation of open source algorithms that allow fast development in the field of Processing and Recognition of images, and 2) works that integrate their functionalities with technological components such as cameras, robots or specialized software. These have the advantage that they improve the images, recognize textures and can distinguish the edges and other main details that appear in the images.

Acknowledgments. We thank the researchers of the Agrarian University of Ecuador for their contribution in the search of the opportune information for the literature review

References

1. Humston, E.M., Knowles, J.D., McShea, A., Synovec, R.E.: Quantitative assessment of moisture damage for cacao bean quality using two-dimensional gas chromatography combined with time-of-flight mass spectrometry and chemometrics. J. Chromatogr. A **1217**, 1963–1970 (2010)
2. Arefi, A., Motlagh, A.M., Khoshroo, A.: Recognition of weed seed species by image processing. J. Food Agric. Environ. **9**, 379–383 (2011)
3. Giraldo-Zuluaga, J.-H., Salazar, A., Daza, J.M.: Semi-Supervised Recognition of the Diploglossus Millepunctatus Lizard Species using Artificial Vision Algorithms. (2016)
4. Szeliski, R.: Computer Vision: Algorithms and Applications. Springer, London (2010)
5. Liu, D., Zeng, X.-A., Sun, D.-W.: Recent developments and applications of hyperspectral imaging for quality evaluation of agricultural products: a review. Crit. Rev. Food Sci. Nutr. **55**, 1744–1757 (2015)
6. Unay, D., Gosselin, B., Kleynen, O., Leemans, V., Destain, M.F., Debeir, O.: Automatic grading of Bi-colored apples by multispectral machine vision. Comput. Electron. Agric. **75**, 204–212 (2011)
7. Muñoz, F.I.I., Comport, A.I.: Point-to-hyperplane RGB-D pose estimation: fusing photometric and geometric measurements. In: IEEE International Conference on Intelligent Robots System 2016 Nov 24–29 (2016)
8. Zhang, B., et al.: Principles, developments and applications of computer vision for external quality inspection of fruits and vegetables: A review. Food Res. Int. **62**, 326–343 (2014)
9. León-Roque, N., Abderrahim, M., Nuñez-Alejos, L., Arribas, S.M., Condezo-Hoyos, L.: Prediction of fermentation index of cocoa beans (Theobroma cacao L.) based on color measurement and artificial neural networks. Talanta **161**, 31–39 (2016)
10. Teye, E., Huang, X.: Novel prediction of total fat content in cocoa beans by FT-NIR Spectroscopy based on effective spectral selection multivariate regression. Food Anal. Methods **8**, 945–953 (2015)
11. Hue, C., et al.: Near infrared spectroscopy as a new tool to determine cocoa fermentation levels through ammonia nitrogen quantification. Food Chem. **148**, 240–245 (2014)
12. Teye, E., et al.: Estimating cocoa bean parameters by FT-NIRS and chemometrics analysis. Food Chem. **176**, 403–410 (2015)
13. Sunoj, S., Igathinathane, C., Visvanathan, R.: Nondestructive determination of cocoa bean quality using FT-NIR spectroscopy. Comput. Electron. Agric. **124**, 234–242 (2016)
14. Armin, L., Adhitya, Y.: Classifying physical morphology of cocoa beans digital images using multiclass ensemble least-squares support vector machine classifying physical morphology of cocoa beans digital images using multiclass ensemble least-squares support vector machine. J. Phys: Conf. Ser. **979**, 10 (2018)
15. Astika, I.W., Solahudin, M., Kurniawan, A., Wulandari, Y.: Determination of cocoa bean quality with image processing and artificial neural network. AFITA 2010 - Comput. Based Data Acquis. Control Agric. **2760**, 6 (2013)
16. Soto, J., Granda, G., Prieto, F., Ipanaque, W., Machacuay, J.: Cocoa bean quality assessment by using hyperspectral images and fuzzy logic techniques. Twelfth Int. Conf. Qual. Control Artif. Vis. **9534**, 1–7 (2015)

17. Ochoa, D., Criollo, R., Liao, W., Cevallos-Cevallos, J., Castro, R., Bayona, O.: Improving the detection of cocoa bean fermentation-related changes using image fusion. Proc. SPIE - Int. Soc. Opt. Eng. **10198**, 1–6 (2017)
18. Ruiz Reyes, J., Soto Bohórquez, J., Ipanaqué Alama, W.: Hyperspectral analysis based anthocyanin index (ARI2) during cocoa bean fermentation process. Proc. - 2015 Asia-Pacific Conf. Comput. Syst. Eng. APCASE 2015 **2**, 169–172 (2015)
19. Veites-campos, S.A., Ramírez-betancour, R.: Identification of cocoa pods with image processing and artificial neural networks. Int. J. Adv. Eng. Manag. Sci. **4**, 510–518 (2018)
20. Hashimoto, J.C., et al.: Quality control of commercial cocoa beans (Theobroma cacao L.) by near-infrared spectroscopy. Food Anal. Methods **11**, 1510–1517 (2018)
21. Ruiz Reyes, J.M., Soto Bohorquez, J., Ipanaque, W.: Evaluation of spectral relation indexes of the Peruvians cocoa beans during fermentation process. IEEE Lat. Am. Trans. **14**, 2862–2867 (2016)
22. Kutsanedzie, F.Y.H., Chen, Q., Hassan, M.M., Yang, M., Sun, H., Rahman, M.H.: Near infrared system coupled chemometric algorithms for enumeration of total fungi count in cocoa beans neat solution. Food Chem. **240**, 231–238 (2018)
23. Huang, X., Teye, E., Sam-Amoah, L.K., Han, F., Yao, L., Tchabo, W.: Rapid measurement of total polyphenols content in cocoa beans by data fusion of NIR spectroscopy and electronic tongue. Anal. Methods **6**, 5008–5015 (2014)
24. Kutsanedzie, F.Y.H., Chen, Q., Sun, H., Cheng, W.: In situ cocoa beans quality grading by near-infrared-chemodyes systems. Anal. Methods **9**, 5455–5463 (2017)
25. Jimenez, J.C., et al.: Differentiation of Ecuadorian National and CCN-51 cocoa beans and their mixtures by computer vision. J. Sci. Food Agric. **98**, 2824–2829 (2018)
26. Bedini, A., Zanolli, V., Zanardi, S., Bersellini, U., Dalcanale, E., Suman, M.: Rapid and simultaneous analysis of xanthines and polyphenols as bitter taste markers in bakery products by FT-NIR spectroscopy. Food Anal. Methods **6**, 17–27 (2013)
27. Lombaert, S.De, Laurent, J., Lehon, M.: Profile of cacao cultivated in Colombia: a study based on standardized methods, indicators of quality and variety. Int. J. Food Nutr. Res. **2**, 1–3 (2018)
28. Hasegawa, R., Hotta, K.: Stacked partial least squares regression for image classification. In: 2015 3rd IAPR Asian Conference on Pattern Recognit, pp. 765–769 (2015)
29. Huang, K., Li, S., Kang, X., Fang, L.: Spectral-Spatial Hyperspectral Image Classification Based on KNN. Sens. Imaging. **17**, 1–13 (2016)
30. Yu, S., Jia, S., Xu, C.: Convolutional neural networks for hyperspectral image classification. Neurocomputing **219**, 88–98 (2017)

Agro-Ecological Zoning of Cacao Cultivation Through Spatial Analysis Methods: A Case Study Taura, Naranjal

Sergio Merchán-Benavides[1](✉), Carlota Delgado-Vera[2], Maritza Aguirre-Munizaga[2], Vanessa Vergara-Lozano[2], Katty Lagos-Ortiz[2], and Tayron Martínez-Carriel[1]

[1] Escuela de Ingeniería Agronómica, Facultad de Ciencias Agrarias, Universidad Agraria del Ecuador, Av. 25 de Julio y Pio Jaramillo, P.O. BOX 09-04-100, Guayaquil, Ecuador
{smerchan, tmartinez}@uagraria.edu.ec

[2] Escuela de Ingeniería en Computación e Informática, Facultad de Ciencias Agrarias, Universidad Agraria del Ecuador, Av. 25 de Julio y Pio Jaramillo, P.O. BOX 09-04-100, Guayaquil, Ecuador
{cdelgado, maguirre, vvergara, klagos}@uagraria.edu.ec

Abstract. The cacao value chain is the third most important after bananas and flowers. The sustainability of natural resources is a subject that should be considered in all aspects of production. In this sense arises the Agro-ecological zoning, a strategy used by the FAO (Food and Agriculture Organization), which helps to characterize zones to determine their agricultural potential. In this work, we propose an approach for Agro-ecological zoning of cacao cultivation through spatial analysis methods. The study area is located in the area of Taura, Parroquia Taura, Canton Naranjal, Province of Guayas. The obtained results show that factors such as the lack of irrigation, poor drainage, poor soil depth, salinity, and sodicity have a negative influence on the cultivation of cacao. Therefore, it is recommended that the management of these factors should be improved with a good drainage network for the evacuation of excess water and elimination of salinity.

Keywords: Agro-ecological zoning · Cacao · Climate · GIS · Soil

1 Introduction

According to studies carried out by CORPEI, 154 545 MT were produced both for the international market (105 000 MT exported) and for the domestic market of 463 787 ha of cacao planted in 2007. The research results showed that 61 073 MT (39%) are produced by small producers, meanwhile, 64 618 MT (42%) are produced by producers of 10–50 ha, and 28 854 MT (19%) are produced by large producers of more than 50 ha [1].

The cacao value chain is the third most important after bananas and flowers. 130 737 MT of cacao beans and 19 966 MT of semi-processed cacao (pasta, liquor, butter) were exported in 2010, reaching a figure record of 150 704 MT which

R. Valencia-García et al. (Eds.): CITAMA 2019, AISC 901, pp. 88–98, 2019.
https://doi.org/10.1007/978-3-030-10728-4_10

represented USD 394 814 627.84 [2]. According to the III National Agricultural Census (2002), there are 463 787 ha of cacao, where about 77 000 ha (17%) are planted in a non-productive stage. It is important to mentioned that more than 50% of the planted area is associated with other fruit or timber species, which means that cacao is a species that contributes to the conservation of natural resources and biodiversity. Also, it protects soil erosion, especially in the foothills (foothills of the mountain range), becoming barriers that stop the dragging of land from the high areas in the winter stage. Cacao is in the hands of 94 855 UPAS (families), 55 499 (59%) are small producers of less than 10 ha, 28 960 UPAS (31%) have 11 and 50 ha, and 10 936 UPAS (11%) are producers of more than 50 ha.

The sustainability of natural resources is a subject that should be considered in all aspects of production [3]. Developing countries carry out plans and strategies to be productive without forcing new areas to be used for agriculture. This ensures that current resources endure and may at some point be better than those we inherit for the next generations. In this sense arises the Agro-ecological zoning, a strategy used by the FAO 1997, which helps to characterize zones to determine their agricultural potential [4].

The Geographic Information Systems allow obtaining more efficient results of the global thematic cartography. These systems are defined as a set of powerful tools to collect, store, retrieve, transform and display spatial data from the real world for particulars objectives [5]. A coordinate system is used to represent the real world, in which the location of an element is given by the latitude and longitude magnitudes in units of degrees, minutes and seconds [6]. GIS can be used in a wide variety of agricultural, livestock [7] and agro-industrial applications, such as crop field management [8], crop rotation monitoring, soil loss projection, and irrigation systems management. In recent years, satellite images have been used to inventory and monitor crops, which allows analyzing the growth, production, predictions of demand and supply, conflicts of land use, among others [9].

This document is structured as follows. Section 2 presents study area, data used in the study, methodology, analysis of climatic variables, and automated information processing with GIS tools. Section 3 describes the discussion of obtained results. Finally, Sect. 4 presents the conclusions and future work.

2 Materials and Methods

This section presents the variables, the data sources and the methods used to achieve the proposed objective.

2.1 Study Area

The study area is located in the area of Taura, Parroquia Taura, Canton Naranjal, Province of Guayas which has an area of 933 087 km^2 and a perimeter of 154 84 km. The quadrangle that limits the study area has the following UTM coordinates: A (630000, 9758000); B (662000, 9758000); C (662000, 9743000); D (630000, 9743000).

The ecological conditions of the area are an average temperature of 25.4 °C (maximum up to 36 °C and minimum up to 15 °C), relative humidity 88%, annual rainfall of 829.4 mm, and brightness of 1004 h/year. The canton has the following limits: southwest with the province of Guayas, north with the cantons Durán and El Triunfo, south to the Balao Canton, east with the provinces of Canar and Azuay, and west to the Gulf of Guayaquil.

2.2 Data

We based on the environmental requirements of the crops and the characteristics of the land for agro-ecological zoning. The temperature and precipitation are critical edafo-climáticos factors for the cultivation of cacao. Cacao is a plant that grows in shade in its natural habitat; however, technified irrigation systems allow exploitation at full exposure.

In this work, The Penman - Monteith method modified by FAO [10] was used to calculate the evapotranspiration. The water balance was made for the study area considering the following variables: Potential Evapotranspiration (PET), average rainfall of the selected series, and water storage capacity in the soil. It is important to consider the geomorphological requirements of the study area because it is a plain relief with very low slope and it is in the typical floodplain of the large river valleys (Río Guayas). Due to this fact, it is susceptible to the periodic floods that occur in the sector. In addition, another factor is the large flows from the drainage of the surrounding watersheds. Table 1 present the adequate climatic and edaphological requirements of cacao cultivation according to the Ministry of Agriculture.

Table 1. Climatic and edaphological requirements of cacao cultivation (Theobroma cacao L.)

Parameter	Required values
Altitude	Less than 700 m above sea-level
Climate	Warm temperate or tropical
Precipitation	More than 1500 mm per year
Temperature	22–28 °C
Lightness	2.74 h/day
RH	60–90%
Wind speed	Up 2 m/s
Photosynthetic assimilation	C3
Depth	More than 50 cm
Texture	Clay loam - sandy loam
Slope	Less than 30%
pH	6.5–7.5
Type	Soils rich in organic matter (>2%), well aired
Life zones	Tropical wet forest (Holdridge)

Regarding the composition and structure of the land, the study area is located on non-consolidated deposits of alluvial origin. Its superficial covering is gray or brown clay with a thickness that varies from 5 to 10 meters approximately according to the lithological records coming from the perforations made in the area.

In order to carry out the calculation of the water balance, two types of information were used: Potential Evapotranspiration and Precipitation. All of them obtained on a multi-year monthly basis. In most cases, they cover a period of 28 years (1980–2008) corresponding to 80% of the total number of stations taken as a reference. It allows greater reliability and accuracy in all calculations of the water balance, since precipitation it is the only source of moisture available in the soil. The Penman equation allows predicting evapotranspiration, in cold and humid zones, as well as in hot and arid zones. The formula is presented below:

$$\boldsymbol{ET}_0 = (\boldsymbol{c}) * [\boldsymbol{W} * (\boldsymbol{Rn}) + (1 - \boldsymbol{W}) + [\boldsymbol{f}(\boldsymbol{u})] * (\boldsymbol{ea} - \boldsymbol{ed})] \quad (1)$$

where

ET_0 = Evapotranspiration in mm/day
(c) = Adjustment factor for Penman
W = Penman Weighting Factor
Rn = Total net radiation in mm/day
f (u) = Wind function
ea = Water vapor pressure at saturation in mbar
ed = Pressure of water vapor to the environment in mbar

The vegetative period was obtained by a simple method that compares P and PET. The initial date is obtained through P = 0.5 * Ep and the final date with the same expression and the consumption of 100 mm remaining from the previous month [11].

2.3 Methodology

The proposed methodology was developed following the model proposed by FAO [4], which consists of collecting climatic information, soil and crop requirements from various bibliographical sources. This information allows creating a database or inventory of climate, soil, and agro-ecological requirements of the crop. This source allows users to manage the geographic information systems in an orderly and effective manner, as well as, obtain the best results in the zoning process (see Fig. 1).

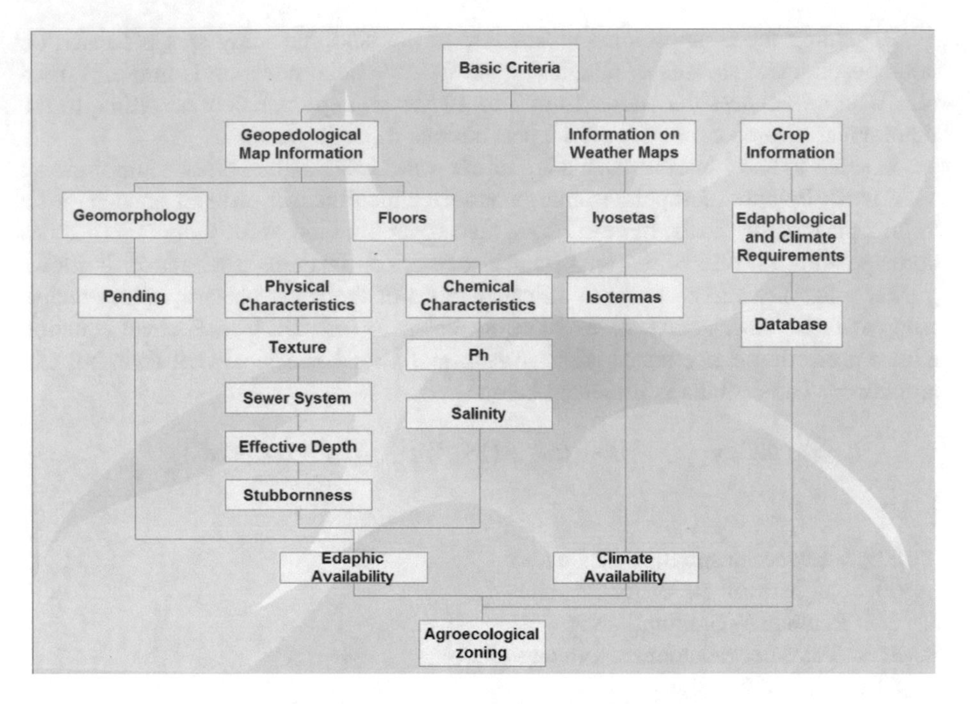

Fig. 1. Agro-ecological zoning scheme

2.4 Analysis of Climatic Variables

The climatic variables obtained through the National Institute of Meteorology and Hydrology at the "Banatel" Hacienda of the Taura Parish were analyzed using the Penman equation [12]. This study allows knowing the relationship and pattern that climatic variables have within the analyzed area. The climatic variables obtained are evaporation, heliofanía, relative humidity, dew, precipitation, average temperature, minimum temperature, and maximum temperature.

Cacao is a plant that grows in shade in its natural habitat. Its lowest annual average temperature is 21 °C according to the FAO. As can be seen in Fig. 2, the average temperature has been optimal to produce cacao from 2008 to 2015.

Figure 3 shows that within the zone, there is much more rainfall than evaporations. This type of scenario is the most recommended since the cacao needs to be irrigated and not be harvested in arid soils.

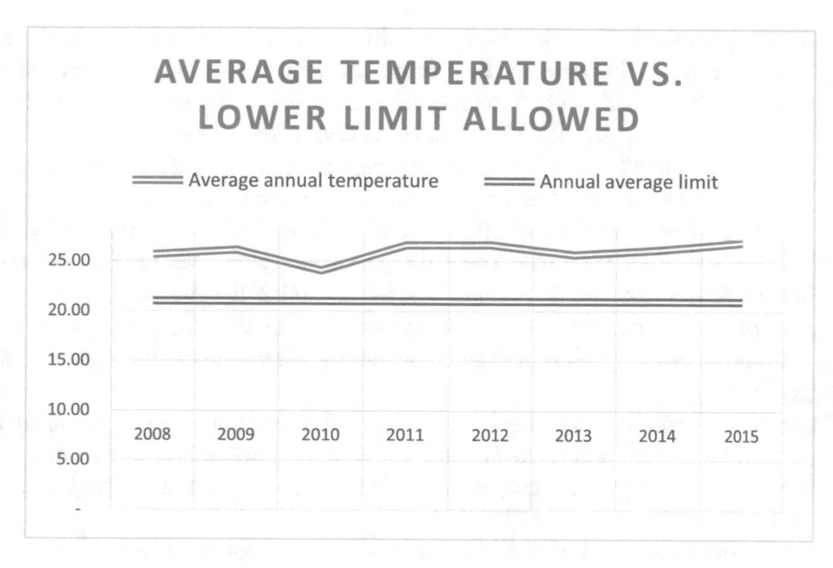

Fig. 2. Comparison of the annual average temperature Vs. lower limit

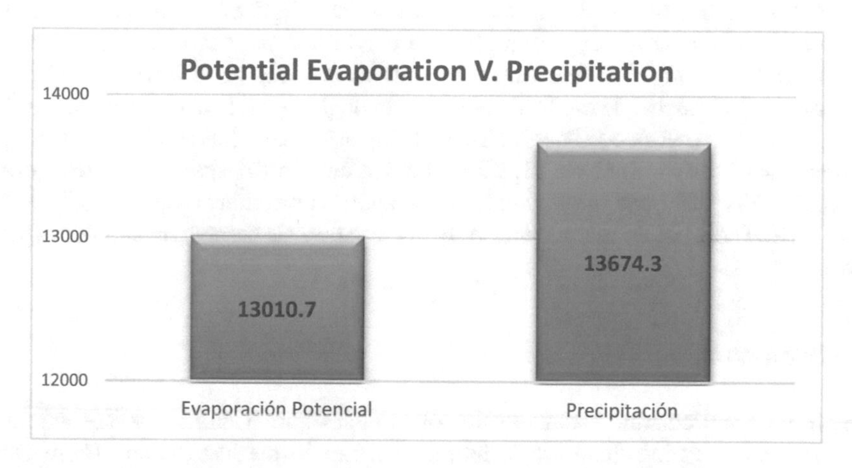

Fig. 3. Compassion of potential evaporation Vs. precipitation

2.5 Automated Information Processing with GIS Tools

We rely on Geographical Information Systems to establish relationships between the different parameters and the agro-ecological zoning, which represents an irreplaceable tool to the use of superposition cartographic methods in the selected natural elements.

The FAO methodology for agro-ecological zoning (AEZ) was designed for computers, using Geographic Information Systems (GIS). It involves the combination of layers of spatial information to define zones. The most advanced research of AEZ incorporates a series of databases, linked to a GIS and related to computer models, which have multiple potential applications in the management of natural resources and the land-use planning. The AEZ study develops an integrated package that can be easily adapted to the conditions and circumstances of any other place. Specifically, the integrated system responds to two main components: (1) a database of land resources, and (2) a set of models, mainly with data collection taken on the land.

The database was obtained by combining various layers of information (maps and tables) on the physical elements of the rural environment, such as soil, relief, and climate. The models were formulated to calculate the aptitude and productivity of the lands, establishing their ideal use. The AEZ provides a global framework to evaluate and plan the land resources. It is important to determine the physical, chemical and biological properties through the land sampling and the laboratory analysis work to create a database properly georeferenced considering a data table that it is used to agro-ecological zoning.

On the other hand, the crop requirements were also examined by the spatial analysis software and they were related to the soil and climate information of the study area. The climatic adaptability of the crop was analyzed comparing the physiological aspects of the plant to produce biomass [13].

Several criteria were established for the cultivation of cacao (Theobroma cacao L) as a result of the homologation of the different parameters of climate and soil variables [14]. Eight classes of the agrological capacity of the land were obtained (see Fig. 4): 1) Class I. It includes good lands, from any point of view (light green color). 2) Class II. It is formed by good lands, but certain physical conditions make it not as rich as those of class I (yellow color). 3) Class III. It includes moderately good lands for cultivation (red color). 4) Class IV. These lands are good enough for vertical cultivation, handling it carefully (blue color). 5) Class V. These lands are almost flat and are not subject to erosion (dark brown). 6) Class VI. Unproductive land in the agricultural field (orange color). 7) Class VII. Land unsuitable for cultivation with serious limitations (brown). 8) Class VIII. These lands are suitable only for hunting or recreation purposes (purple color).

3 Discussion

The climatic and edaphic characteristics of the Parroquia Taura show that within the agro-ecological zoning there are 3 delimited zones to produce cacao. These results directly demonstrate dependence on soil and climatic factors that in some cases limit the sustainable improvement of production. These facts agree with Odum in 1984 [15], who describes that ecological factors intervene in the production and normal development of crops. Furthermore, the existing limitations in the two production levels are generated by soil aspects such as drainage, soil depth, salinity and groundwater levels [16]. Therefore, it is necessary to identify areas that can generate sustainability in the production of crops.

The areas that present favorable conditions to produce cacao must have characteristics such as: 1) flat or undulating land, 2) depth must be in the acceptable range to root exploration and anchoring, 3) stoniness must not influence in the land, and 4) a good branch design to dislodge the surplus water.

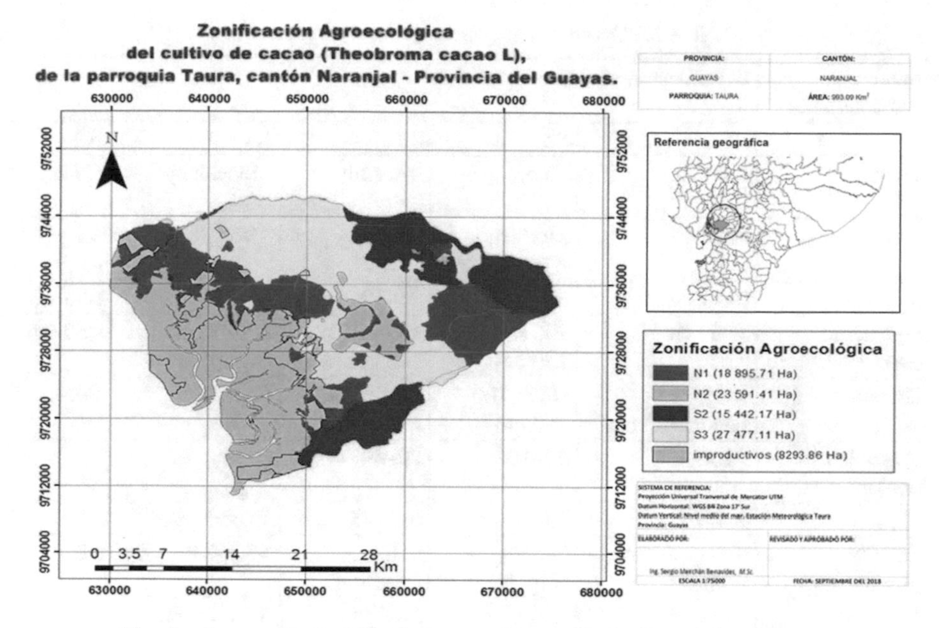

Fig. 4. Agro-ecological Zoning map of cacao in the Parroquia of Taura

Conversely, the unproductive areas of the Parroquia Taura correspond to those areas in which the slope exceeds 30%, which correspond to natural bodies of water, urban settlements, rocky outcrops, sandbanks, littoral cords, and ecological reserves. These lands are inadequate and uneconomic because of the number of resources that should be applied, and because they are protected areas of fauna and flora.

Irrigation scheduling and calibration is an essential element for the long-term stability of cacao production [17]. In [18], authors indicate that the drought occurrence frequency or water scarcity during the annual cacao cycle has key effects on the expected production. The size of the grains will be smaller and the phase of vegetative growth of the plant at the end of the harvest will be severely affected, reducing its photosynthetic capacity.

The results presented in Table 2 show that the climate with its different variables is not a limitation in the area analyzed to the harvest of cacao, because parameters are within those accepted or tolerable by cacao. Specifically, tropical climates are ideal to develop of perennial grain crops [19, 20]. The results also show that Parroquia Taura has an area with very good characteristics to harvest cacao. It represents 29.32% of its surface which has a clay loam texture. Alluvial soil is one of the best soils for cultivation since cacao a is a perennial crop that needs deep soils (more than 1 m), with medium texture [21, 22].

Table 2. Cacao soil requirements and characteristics

Land characteristics	Class degree of limitation and rating				
	S1	S2	S3	N1	N2
Texture	Medium	Fine	Very fine	Moderately coarse	Coarse
Soil color	Dark	Mottled	Mottled	Clear	Very clear
Slope (%)	0–5	6–16	16–30	30–50	>50
Soil depth cm)	>200 cm	200–150 cm	150–100 cm	100–50 cm	<50 cm
Drainage	Very good >150 cm	Good 100–150 cm	Moderate	Imperfect	Poor
Water table (cm)	>100	50–100	20–50	0–20	"
pH	6.5–7.5	7.6	5.5–4.5	Waterlogged"	
Salinity (mmhos/cm)	0–2 (without salinity)	2–4 (Light)	4–8 (medium)	>7.7–7.9	>8
Toxicity (Al) (CO3Ca)	No	Light	Medium	8–16 (high)	>16 (very high)
M.O. %	>10%	4–10%	2–4%	High	Very high
Stoniness	<10 without stoniness	10–25 little stoniness	<10 without stoniness	1–2%	<1%

4 Conclusions and Future Work

In this work, we proposed an approach for agro-ecological zoning of cacao cultivation through spatial analysis methods. The spatial analysis tool used allows taking decisions in a better way in the agricultural sector. However, we found as a limitation the lack of continuous meteorological data of all the agricultural areas of the country, which can be improved by taking reference data from global databases.

With a zoning of all the crops in Ecuador, climate risk is potentially improved for small farmers, which significantly improves the welfare of agricultural households.

The data generated from this research allow us to propose predictive models of crop behavior. Also, the proposed methodology can be classified as bio-economic, which allows optimizing natural resources during the harvest. In addition, the methodology can analyze the effects of technological change in the analysis of edaphological information to promoting economic benefit and sustainability.

References

1. Secretaría Técnica del Comité Interinstitucional para el Cambio de la Matriz Productiva-Vicepresidencia del Ecuador, Cepal: Diagnóstico de la Cadena Productiva del Cacao en el Ecuador. Ecuador (2014)
2. ANECACAO busca reconocimiento de Ecuador en el mundo (2015)
3. Wezel, A., Casagrande, M., Celette, F., Vian, J.F., Ferrer, A., Peigné, J.: Agroecological practices for sustainable agriculture. A review. Agron. Sustain. Dev. **34**(1), 1–20 (2014)
4. FAO: Agro-ecological zoning guidelines. FAO Soils Bull. **76,** 3–5 (1996)
5. Uyaguari, A., Espinosa-gallardo, E., Santiago, P.J., Espinel, P., Alberto, F., Calder, C.: Open Source Web Software Architecture Components for Geographic Information Systems in the Last 5 Years: A Systematic Mapping Study, p. 721 (2018)
6. Burrough, P.A., McDonnell, R., McDonnell, R.A., Lloyd, C.D.: Principles of Geographical Information Systems. OUP, Oxford (2015)
7. Höhn, J., Lehtonen, E., Rasi, S., Rintala, J.: A geographical information system (GIS) based methodology for determination of potential biomasses and sites for biogas plants in southern Finland. Appl. Energy **113**, 1–10 (2014)
8. Marsiglia Fuentes, R.M., Torregroza Fuentes, E., Quintana, S.E., García Zapateiro, L.A.: Application of geographic information systems for characterization of preharvest and postharvest factors of squash (cucurbita sp.) in Bolívar Department, Colombia. Indian J. Sci. Technol. **11**, 1–10 (2018)
9. Ottinger, M., Clauss, K., Kuenzer, C.: Aquaculture: relevance, distribution, impacts and spatial assessments – a review. Ocean Coast. Manag. **119**, 244–266 (2016)
10. Pereira, L.S., Allen, R.G., Smith, M., Raes, D.: Crop evapotranspiration estimation with FAO56: past and future. Agric. Water Manag. **147**, 4–20 (2015)
11. Food and Agriculture Organization for the United Nations: FAO Statistical Yearbook 2013: World Food and Agriculture. Roma. www.fao.org/publications (2013)
12. Córdova, M., Carrillo-Rojas, G., Crespo, P., Wilcox, B., Célleri, R.: Evaluation of the Penman-Monteith (FAO 56 PM) method for calculating reference evapotranspiration using limited data. Mt. Res. Dev. **35**, 230–239 (2015)
13. Wartenberg, A.C., Blaser, W.J., Gattinger, A., Roshetko, J.M., Noordwijk, M.Van, Six, J.: Does shade tree diversity increase soil fertility in cocoa plantations? Agric. Ecosyst. Environ. **248**, 190–199 (2017)
14. Snoeck, D., Koko, L., Joffre, J., Bastide, P., Jagoret, P.: Cacao nutrition and fertilization. In: Lichtfouse, E. (ed.) Sustainable Agriculture Reviews, vol. 19, pp. 155–202. Springer International Publishing, Cham (2016)
15. Altieri, M.A.: Agroecology: the Science of Sustainable Agriculture. CRC Press, Boca Raton
16. Zech, W.: Geology and Soils. In: Pancel, L., Köhl, M. (eds.) Tropical Forestry Handbook, pp. 1–191. Springer, Berlin (2016)
17. Mommer, L.: The water relations in cacao (Theobroma cacao L.): Modelling root growth and evapotranspiration. 57 (1999)
18. Carr, M.K.V., Lockwood, G.: The water relations and irrigation requirements of cocoa (Theobroma cacao l.): a review. Exp. Agric. **47**, 653–676 (2011)
19. Corlley, R.: Potential productivity of tropical perenial crops. Exp Agric. Exp. Agric. **19**, 217–237 (1983)
20. Salazar, O.V., Ramos-Martín, J., Lomas, P.L.: Livelihood sustainability assessment of coffee and cocoa producers in the Amazon region of Ecuador using household types. J. Rural Stud. **62**, 1–9 (2018)

21. Willer, H., Lernoud, J.: The World of Organic Agriculture 2016: Statistics and Emerging Trends. (2016)
22. Claver-Cortés, E., González Illescas, M., Zaragoza-Sáez, P.C., Vargas Jiménez, M.: Knowledge management in cocoa artisanal firms. The case of "El Oro" province (Ecuador). Przedsiębiorstwo i Reg. **9**, 131–141 (2017)

PestDetect: Pest Recognition Using Convolutional Neural Network

Federico Murcia Labaña, Alberto Ruiz, and Francisco García-Sánchez(✉)

Faculty of Computer Science, Department of Informatics and Systems, University of Murcia, 30100 Murcia, Spain
{federico.murcia,aruiz,frgarcia}@um.es

Abstract. Agriculture is a strategic sector in many regions around the world. In those regions where water scarcity is an endemic problem, crops tend to suffer hydric stress which make them prone to suffer from pests and diseases. Thus, periodic checks to detect those pests are crucial to prevent and act upon them on early stages. Portable smart devices like phone mobiles or tablets offer Internet connectivity and camera devices. These two properties make them a potential tool that can be used for this work: to make an in situ early detection of the pest or disease that could help to reduce the negative impact of these on the affected crop and so minimize economic loss. In this work we propose an application prototype that can issue a diagnosis and its related treatment from a photograph of an affected crop taken by the user anytime/anywhere. This is achieved by using a combination of different technologies such as Convolutional Neural Networks and REST services, among others. The first tests with a reduced set of crops and diseases resulted in an accuracy over 90%.

Keywords: Convolutional neural network · Diseases classifier
Plant pest recognition

1 Introduction

New techniques have been applied to farming to maximize production and minimize costs while keeping quality standards. In those places where hydric resources are limited, crops are prone to pests and diseases [1]. Certainly, all crops are to suffer from pests or diseases during production. Thus, it is mandatory to perform periodic controls as a prevention mechanism. Without these, farming would be unsuitable for the actual population demand. Not detecting crop pests and diseases at an early stage may lead to crop loss and serious financial problems for the farmer.

Nowadays, some places have implemented sophisticated resources monitoring and management systems that have allowed to increase production despite of water scarcity. For example, the Water Technology Centre[1] (CETAQUA), a leading technology center in water cycle with offices worldwide, is currently implementing Machine

[1] http://www.cetaqua.com/en.

R. Valencia-García et al. (Eds.): CITAMA 2019, AISC 901, pp. 99–108, 2019.
https://doi.org/10.1007/978-3-030-10728-4_11

Learning solutions for monitoring hydric resources intended for crops irrigation [2]. Information Technologies (IT) are changing almost every part of our society at rapid pace. As they become more accessible to users, even the most conservative sectors such as the agriculture sector benefit from the opportunities provided by Internet and other Information and Communications Technologies (ICT). Information is everywhere, and it can be used to create tools that may help improve any discipline. Applications that only were possible to execute in a Personal Computer (PC) some years ago, can be now executed in any portable device like a smartphone or tablet anytime and anywhere. Today, these devices can be used as low-cost tools to help determine the presence of pests or diseases in plants and classify them.

In this work, we propose a machine learning-based framework to assist farmers and agricultural holding managers in identifying and classifying the pests and diseases affecting their crops. It consists of a responsive web application through which users can send images of the affected areas of the plant. These images are processed by a pre-trained Convolutional Neural Network [3], which identifies and classifies the condition if successful, providing information about the most convenient treatments to apply. The tests performed on a reduced set of pests present in citric crops (e.g., lemon/orange tree) resulted in an accuracy of over 90%. With all, two main goals are achieved: (i) to assist in the early identification of crop pests and diseases, and (ii) to suggest preventive treatments so as to prevent the condition from spreading, thus avoiding greater damages to crops and limiting financial losses.

The rest of this paper is structured as follows. Section 2 reviews related work by other authors. Section 3 presents the proposed framework to detect pests and diseases from the photos taken in the field. The proof-of-concept implementation of the framework is described in Sect. 4. Finally, conclusions and future work are put forward in Sect. 5.

2 Related Work

A number of different related approaches have been recently published in indexed peer reviewed journals [4–6]. The one most closely related to the framework proposed here is [4], which features a Multi-resolution mobile vision system for plant leaf disease diagnosis. It consists of a client-server architecture that rests on a web application providing services concerned with disease classification on crop leaves. In that work, the classification approach to detect diseases is as follows:

1. Take a photograph of the affected leaf with a portable smart device.
2. Preprocess the photograph on the smart device, applying some techniques like color space conversion, normalization, noise filtering and image segmentation with the objective of isolating the affected zone.
3. Once the image has been preprocessed, it is sent to the server.
4. Once the image is in the server, a combination of transformations is made on the image to (i) reduce the problem dimensionality by applying feature extraction techniques, and (ii) achieve a resolution changes-tolerant solution. At this phase, Gabor Transform (GT) satisfies the multi-scale feature extraction for real objects

while the Wavelet Transform (WT) fulfills the requirement of multi-resolution features, and Gray Level Co-occurrence Matrix (GLCM) opens a new dimension in pattern recognition.

5. Finally, the final processed image serves as input to five classification algorithms, namely, K-Nearest Neighbors (k-NN), Linear Regression, K-Star, Radial Basis Function Support Vector Machine (RBF SVM), and M5 model tree (M5P).

The results of each of these classifiers as stated by the authors are shown in Table 1.

Table 1. Accuracy comparison of different texture feature extraction algorithms and classifiers.

Classifiers	Accuracy (%)		
	WT	GWT	GWT + GLCM
M5P	78.47 ± 3.16	81.23 ± 1.86	82.44 ± 3.50
Linear regression	86.27 ± 5.11	85.97 ± 5.11	88.27 ± 5.11
k-Star (k*)	78.25 ± 3.67	79.00 ± 3.83	81.00 ± 4.40
RBF SVM	80.44 ± 2.12	85.69 ± 1.56	88.56 ± 3.51
k-NN	88.00 ± 4.89	90.15 ± 3.60	93.50 ± 2.89

The main advantages of this approach are: (i) the photograph is preprocessed, isolating the affected zone and so only affected zones will be analyzed; (ii) different transformations are considered to allow classification algorithms to be tolerant to different image resolutions and to improve the classification rate; and (iii) it achieves consistent results among algorithms. However, also some drawbacks can be listed. First, the image preprocessing stage takes place in a mobile device; thus, the execution time depends on the device hardware. Second, the photograph provided by the client should be processed in the server to avoid different image resolutions. Finally, the application has been developed to run on Android-based devices, so it is not possible to use the app from other devices.

The solution proposed in this manuscript, namely, "PestDetect", make use of Convolutional Neural Network algorithms to classify images of plant pests and diseases. The proof-of-concept implementation has been developed in the form of a responsive web application thus allowing access from multiple devices.

3 Proposed Framework

The purpose of this work is to design a framework that farmers could use as an early stage pest/disease detector that prevents the harmful condition to further spread. The functional architecture of the PestDetect framework is depicted in Fig. 1.

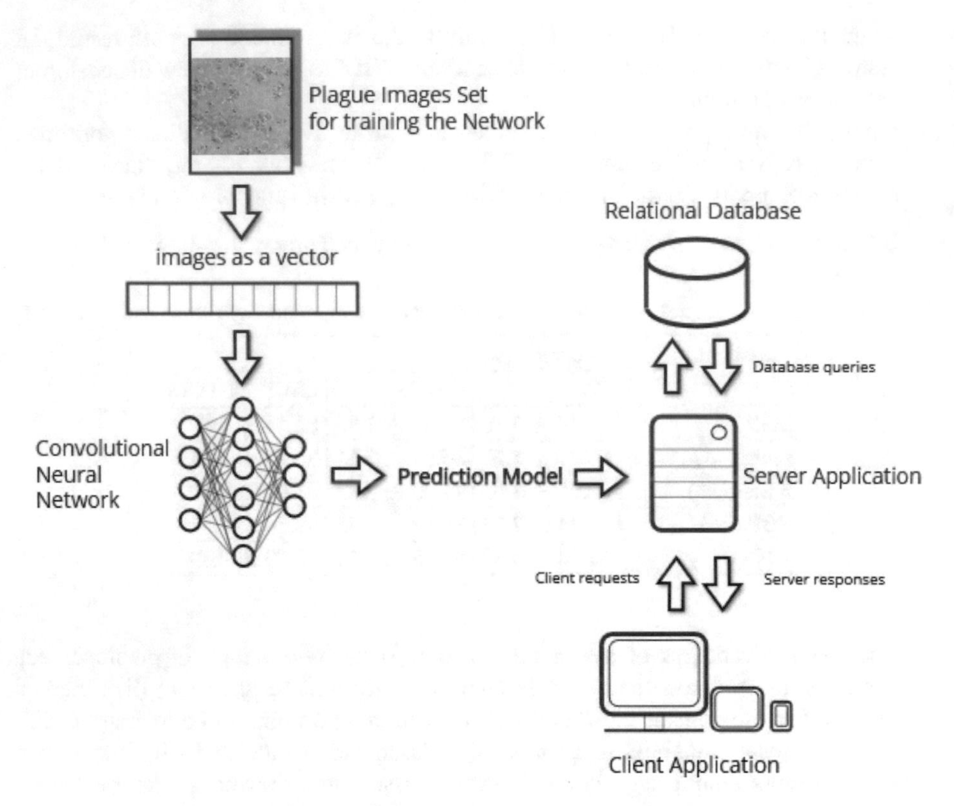

Fig. 1. PestDetect proposed architecture.

3.1 The Convolutional Neural Network

Convolutional Neural Networks allows to reduce one of the principal problems of the Deep Neural Networks when the input is an image: scalability. A traditional Neural Network is built with a set of different types of interconnected layers. The first layer will have as many inputs (neurons) as input parameters. Each neuron of this layer is connected to each neuron of the next layer. This type of layer is known as "*fully connected layer*". Several fully connected layers create a network. The problem is that this scheme is unsustainable because of the exponential parameters growth. For example, for a computer, a grey scale image is a set of numbers disposed on a single matrix that codifies grey pixel values. So, a 6 megapixels image will take 6 million input parameters. If the first inner layer has 100000 neurons (with a bias input each), the number of parameters to compute will be $6 * 10^6 * (100000 + 100000) = 1200 * 10^9$ parameters. Consequently, this type of network does not scale well. Convolutional Neural Networks, on the other hand, solve this scalability problem by using the following two operations [7]:

1. Convolution operation: each neuron implementing a convolution function is capable of determining local features on an image (see Fig. 2 [left]) by sweeping the

image with convolutional filters. Besides, it brings two interesting properties: (i) the learned patterns are translation invariant, that is, each pattern can be detected in any location within an image, and (ii) it can create spatial hierarchy patterns (see Fig. 2 [right]). A convolution layer is built with neurons that implement the convolution function. This type of layer is defined by two parameters, namely, the size of the convolution filters, and the depth of the feature map[2]

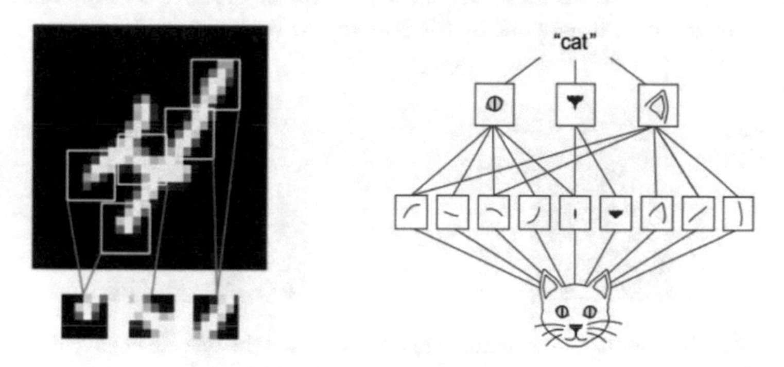

Fig. 2. Images can be fragmented in local patterns like borders, textures, etc.

2. Max-pooling operation: this operation reduces the dimensionality of the feature maps extracted by the convolution layer, obtaining the maximum value of each one.

The configuration of a Convolution Neural Network consists of alternate convolution layers and max-pooling layers. The penultimate layer would be a fully connected type and the last layer, since we are dealing with a classification problem, would have as many neurons as classes to classify. Our proof-of-concept prototypical implementation as described in Sect. 4 only takes into consideration three classes, that is, three neurons containing the probability of been plague type 1, plague type 2 and not a plague.

3.2 Dataset

To achieve the classification goal, it is first necessary to collect information about the crop pests and diseases under consideration. During this stage it is required to (i) collect a large number of images related to each of the relevant conditions to recognize, and (ii) gather knowledge about the appropriate treatments conceived to deal with those pests and diseases. Once that information is available, a study of all the visual information is made to select the pests and diseases from which enough material is available. In the case of our proof-of-concept implementation (Sect. 4), the initial set of pests and diseases includes the following conditions:

---

[2] A feature map is the name given to a tensor (i.e., a vector/matrix container) where an image is stored.

1. Scirtothips
2. Pezothrips
3. Parlatoria pergandii
4. Aonidiella aurantii

Parlatoria pergandii (Fig. 3 [left]) and *Aoniediella aurantii* (Fig. 3 [right]) were finally the selected pests due to the number of images available and the similarity between them, which make them the most suitable in order to test the classification capabilities of the neural network that it is going to be built.

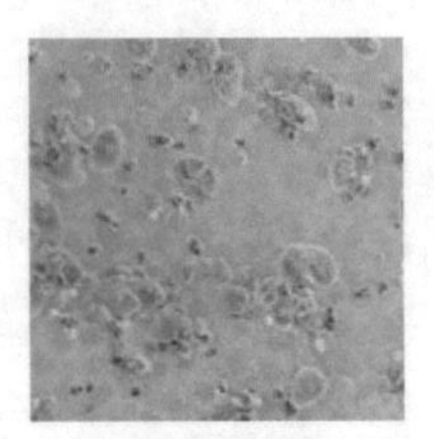
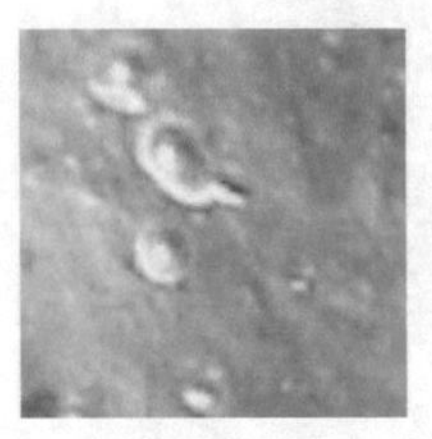

Fig. 3. Left-hand Parlatoria pergandii, right-hand Aoniediella aurantii.

The main problem is that the number of photographs of each plague were relatively low (between 40/50 photographs per plague) and the resolution varies among them (between $3072 * 2304$ pixels and $1417 * 1013$ pixels). The total number of images to train and validate the neural network is insufficient, and was therefore necessary to obtain more images from the initial set of photographs by following this process:

1. For each photograph, obtain a sub-image where the plague is present.
2. For each sub-image, create new images by applying operations of rotation (steps of 60 degrees). With this single step four new images are created, additional to the original one (five images in total).
3. For each image created in step 2, apply a blur effect.
4. For each image created in step 2, apply a noise effect.

At the end of the process, about 1000 images for each plague were available, that is, 2000 images in total. Also, in order to avoid the problem of different image resolutions, each image is chunked by $150 * 150$ pixels. With this the image resolution is unified to feed the neural network ($150\ width * 150\ height\ * 3\ color\ channels = 67.500\ input\ parameters$) without making the network too big, as it will be trained in a domestic PC[3]. Chunking the 2000 images in $150 * 150$ pixels results in about 3000 images for each plague (about 6000 images in total). That is a considerable amount of visual information that allows to train and validate the neural network avoiding overfitting.

[3] Domestic PC Hardware: Intel Core i7 4970 k – 4 GHz CPU, 8 GB RAM, Nvidia Geforce GTX 1060 – 6 GB Video RAM. 1280 CUDA Cores.

3.3 Structure of the Neural Network

Keras/Tensorflow High-level API [8, 9] simplifies the creation of neural networks and the feeding process, which consists in the creation of two sets of images. There will be two training sets and two validation sets for each pest (*Parlatoria pergandii* and *Aoniediella aurantii*). Once the sets are ready, they can feed the network to train it and, later, to validate the already trained network.

The training and validation stages work as follows. An epoch (i.e., one forward pass and one backward pass of all the training examples) has several iterations, and in our case, we run 50 epochs with 100 iterations each. Each iteration creates a random subset of 64 images selected from the training set, whereas the validation subset contains 32 images. As random subsets are created, some random operations (e.g., rotations, shift, zoom, flip) are applied to each image to prevent overfitting. With all, the structure of the proposed Convolutional Neural Network is depicted in Table 2.

Table 2. Convolutional neural network structure.

Layer number	Layer type	Neuron Nº	Activation function
1	Convolution	32	ReLU[4]
2	Max-poolling	32	–
3	Convolution	64	ReLU
4	Max-poolling	64	–
5	Convolution	128	ReLU
6	Max-poolling	128	–
7	Convolution	128	ReLU
8	Max-poolling	128	–
9	Flatten	128	
10	Dropout	128	–
11	Fully connected	64	ReLU
12	Fully connected	3	SoftMax[5]

Once the structure is defined, an optimizer, a loss function and a metric to optimize must be specified. These are AdamMax [10], Categorical Cross-Entropy [11] and Accuracy, respectively.

3.4 Training and Validation Results

The results obtained by considering the above-mentioned configuration are shown in Table 3. With these encouraging results, the model was exported to the server application to be able to make predictions.

[4] ReLU: Rectified Linear Units. This function is defined as $f(x) = max(x, 0)$.

[5] SoftMax adjust values between [0, 1]. These values are related with the probability of each class.

Table 3. Results of training/validation.

Epoch	Iteration	Training/validation loss	Training/validation accuracy
50	100	0.1287/0.1986	0.9480/0.9100

4 Proof-of-Concept Implementation

A prototypical responsive web application has been developed that take advantage of the classification model built as described above to provide farmers with a service to detect the pests and diseases affecting their crops. The Bootstrap library [12] has been used in the front-end side of the application (see Fig. 4) to make the application adjust to different devices. The application features a free subscription platform that allows farmers to use the provided service. Once users are registered in the system, they can login and use the diagnosis service by simply uploading an image region of interest (ROI) which is affected by a pest or disease. If the diagnosis is successful, a convenient treatment for the plague will be shown. To classify pictured condition, the Keras/Tensorflow model built as described in Sect. 3 is used.

The client-server communication is made via REST services [13], which simplifies the way the information is exchanged between the front-end (Ajax client) and the backend (Jersey-based REST services). The services and business model classes, which have been implemented in Java [14], are deployed in an Apache Tomcat server [15] with a MySQL database [16]. The homepage of the web application is shown in Fig. 4.

Fig. 4. Application main screen (in Spanish)

Figure 5 depicts a diagnosis request. For this, the user first uploads a photograph of the affected crop. Then, the user has to select an excerpt of the image (i.e., the area in which the effects of the pest or disease are patent and clear) to be sent to the server for diagnosis (note the red square over the image, Fig. 5 [left]). Finally, if the diagnosis was successful, it is shown to the user, along with the proposed treatment (see Fig. 5 [right]).

Fig. 5. 'Send to diagnosis' screen [left]; 'Issued diagnosis' screen [right] (in Spanish)

5 Conclusions and Future Work

Nowadays, computer vision is becoming a very important research topic because of its great applicability and usefulness in many and heterogeneous areas. All kinds of applications make use of this technique for tasks such as facial recognition, autonomous driving, automatized disease diagnostics, photography, etc. In the agricultural field, the opportunities related with the use of computer vision solutions are enormous. In this work, we explore its use to detect pests and diseases in plants at an early stage so as to prevent them to spread and provide relevant treatments. So far, the preliminary tests have resulted in an accuracy over 90% by considering a very limited number of pests. As a first prototype, its functionality is very basic. It can only classify between two plagues.

We plan to extend and improve this work in the following aspects:

- Expert User features. In case an image is not classified over a trust threshold (75%), it is shown to expert users for them to provide an accurate diagnosis. These images, along with others that can be provided by expert users themselves, can be used to re-train the network.
- Make the Convolutional Neural Network Multi-classifier. The Convolutional Neural Network will be trained to classify multiple types of plagues. This only needs an appropriate set of training images, as the proposed network architecture and preprocessing stages can be reused with minimal changes.

– Use of other indicators. A textual description of the pest/disease effects and other environmental parameters (e.g., weather conditions) can help improve the diagnosis accuracy.

Acknowledgements: Thanks to Mr. Alfonso Lucas Espadas (Plant Health Service, Region of Murcia, Spain) for providing the full pool of images and for his advice. This work was partially supported by the Spanish MINECO, as well as European Commission FEDER funds, under grant TIN2015-66972-C5-3-R.

References

1. Ayres, P.G.: Water relations of diseased plants. In: Water and Plant Disease, pp. 1–60. Elsevier (1978)
2. CETAQUA: Artificial intelligence for agricultural water demand forecasting in South-Eastern Spain. http://www.cetaqua.com/en/press-room/new/526/artificial-intelligence-for-agricultural-water-demand-forecasting-in-south-eastern-spain (2018). Accessed 30 Sept 2018
3. Khan, S., Rahmani, H., Shah, S.A.A., Bennamoun, M.: A guide to convolutional neural networks for computer vision. Synth. Lect. Comput. Vis. **8**, 1–207 (2018)
4. Prasad, S., Peddoju, S.K., Ghosh, D.: Multi-resolution mobile vision system for plant leaf disease diagnosis. Signal Image Video Process. **10**, 379–388 (2016)
5. Iqbal, Z., Khan, M.A., Sharif, M., Shah, J.H., ur Rehman, M.H., Javed, K.: An automated detection and classification of citrus plant diseases using image processing techniques: a review. Comput. Electron. Agric. **153**, 12–32 (2018)
6. Sun, G., Jia, X., Geng, T.: Plant diseases recognition based on image processing technology. J. Electr. Comput. Eng. **2018**, 1–7 (2018)
7. Chollet, F.: Deep Learning with Python. Manning Publications (2017)
8. Keras. https://keras.io/ (2018). Accessed 30 Sept 2018
9. TensorFlow. https://www.tensorflow.org/ (2018). Accessed 30 Sept 2018
10. Kingma, D.P., Ba, J.: Adam: a method for stochastic optimization. In: 3rd International Conference for Learning Representations, San Diego (2015)
11. Hu, K., et al.: Retinal vessel segmentation of color fundus images using multiscale convolutional neural network with an improved cross-entropy loss function. Neurocomputing **309**, 179–191 (2018)
12. Bootstrap. https://getbootstrap.com/ (2018). Accessed 30 Sept 2018
13. Fielding, R.T.: Architectural Styles and the Design of Network-based Software Architectures. University of California, Irvine (2000)
14. Java. https://www.java.com (2018). Accessed 30 Sept 2018
15. Apache Tomcat. http://tomcat.apache.org/ (2018). Accessed 30 Sept 2018
16. MySQL. https://www.mysql.com/ (2018). Accessed 30 Sept 2018

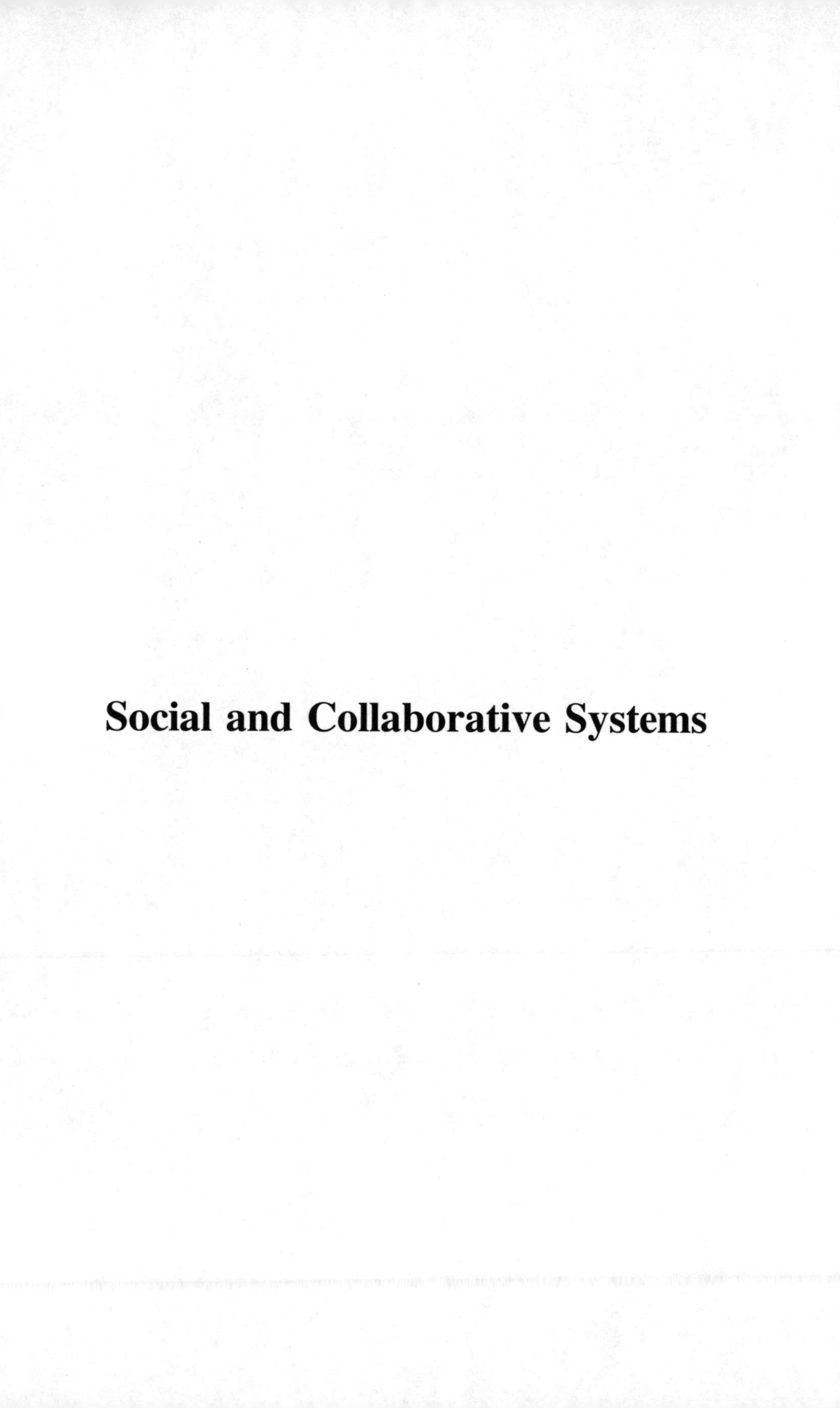

Social and Collaborative Systems

Citizen Science in Agriculture Through ICTs. A Systematic Review

Mitchell Vásquez-Bermúdez(✉), Jorge Hidalgo, Karla Crespo-León, and Jaime Cadena-Iturralde

Agrarian Sciences Faculty, Agrarian University of Ecuador, Avenue 25 de Julio and Pio Jaramillo, Guayaquil, Ecuador
{mvasquez,jhidalgo,kcrespo,jcadena}@uagraria.edu.ec

Abstract. The constant evolution of citizen science in recent years- where the participation of citizens in scientific projects is applied with the use of information technologies- has been able to solve different problems in different areas of knowledge. In the same way, when we deal with the issue of agriculture, together with the support of information technologies, these are applied to increase productivity, automate processes and train farmers to make decisions. There are different processes in agriculture that can be implemented with ICT solutions. This work is focused on carrying out a process of systematic mapping of research which will include a search, study selection, analysis and identification of potential relationships on citizen science in agriculture through ICTs. With the application of systematic mapping, the frequency of indexed publications that has existed on this topic will be evaluated, since no statistical studies on citizen science have been found that account for the number of existing projects on agriculture and technology.

Keywords: Citizen science · Agriculture · ICTs

1 Introduction

In recent years, citizen science has been presented as a new type of scientific production according to Finquelievich and Fischnaller [1]. Citizen science consists of the collaboration, consent and voluntary participation of citizens capable of generating data and information relevant to a specific topic.

The growth of citizen participation in search, data collection, exchange of ideas, analysis, dissemination of information for scientific purposes has grown thanks to the help provided by information technologies. In this way, information technology is applied in different areas of knowledge. In this research we will focus on the area of agriculture, using ICT to increase productivity, improve processes as a means of training, for dissemination of knowledge, as a means to support environmental use, and also to help farmers to make decisions.

According to Pérez, Milla and Mesa [2] ICTs have modified the way of living, the way of seeing the world, since they are used as strategic tools for the development of society. ICTs have contributed to the development of agriculture and are applicable in all agricultural phases and activities [3].

R. Valencia-García et al. (Eds.): CITAMA 2019, AISC 901, pp. 111–121, 2019.
https://doi.org/10.1007/978-3-030-10728-4_12

The objective of this paper is to review the current trend of citizen science as a contribution to agriculture through the use of ICTs, making a systematic mapping in scientific libraries with high impact to obtain results.

2 Systematic Review

Within the processes of agriculture there are several technological tools that can be implemented for their use, in the same way it is important to provide a general perspective on the role of information technologies within agriculture, since with this we could help farmers to optimize processes and operations for the management of resources. Considering what has been previously questioned, a systematic mapping has been carried out with the review approach proposed by Kitchenham and Charters [4]. They carried out a study on a general vision of a research area. It seeks to know the topics discussed. Another work is that of Bazán et al. [5] in which a technological search was carried out in the context of agriculture, finding support systems for decision-making, semantic web, cloud computing, the Internet of things and big data. It points out that the use of ICT has a great effect on agriculture in terms of production and costs.

That is why carrying out a systematic review will allow the identification, evaluation and analyses of several related studies. This work will be divided into three stages, the first stage is to perform an analysis and planning of the search, in the second stage we proceed with the execution of search and review of found articles, and the third stage will be a discussion of results and conclusions.

3 Methodology

3.1 Research Analysis

In the present work, we will apply the methodology proposed by Bazán et al. [5] based on what was proposed by Kitchenham and Charters [4], combining the best references to propose and develop an updated guide.

The systematization of citizen science in agriculture through ICTs was determined, which has led to the following research questions:

- What are the guidelines for carrying out the systematic process of citizen science studies in agriculture through ICTs?
- Where and when were the citizen science studies published in agriculture through ICTs?
- What has been the impact of citizen science and its contribution to agriculture using ICT?

3.2 Search Planning

In order to carry out the search planning we used the recommendation proposed by Kitchenham and Charters [4], for the identification of key words and thus formulate search chains using the PICO method consisting of population, intervention, comparison and results.

The population refers to citizen science that is involved in projects and applications of agriculture. In our context, the population is systematic mapping studies. The intervention refers to the collection obtained by citizen science and the analysis of the data that is supported by the different agricultural collaborators. The comparison was done where a study was carried out relating the different strategies proposed. The results are not considered specific, since the study focuses on empirical studies that evaluate citizen science in agriculture.

The key words for the identification are: citizen science, agriculture and ICTs. They have been grouped together. Certain synonyms have also been considered to formulate a search chain described in the following steps:

Step 1: Determine the scope of the search for citizen science, that is, "citizen science".

Step 2: Determine the search directly related to the intervention, that is, "Agriculture" and "ICTs".

Step 3: Determine the search related to the project, categorization, methods, tools

The search was conducted in the databases of the IEEE Xplore, ACM, Inspec / Compendex and Scopus (ScienceDirect).

Table 1 shows the databases and the search condition.

Table 1. Search conditions

Databases	Search
IEEE	("Citizen science" AND ("Agriculture" OR "agricultural science" AND "information technology" OR "computer applications" OR "Software"))
ACM – Digital Library	"citizen participation" AND ("Agriculture" OR "agricultural science" OR "agronomy study" OR "agronomy") AND ("information technology" OR "Software" OR "technology" OR "computer applications")
Scopus (ScienceDirect)	"citizen science" AND ("Agriculture" OR "agricultural science" OR "agronomical science study" OR "agronomy study" OR "agronomy") AND ("information technology" OR "Software" OR "technology" OR "computer applications")
Inspec/Compendex	("citizen participation" OR "citizen science") AND (("Agriculture" OR "agricultural science" OR "agronomical science study" OR "agronomy study" OR "agronomy") AND ("information technology" OR "Software" OR "technology" OR "computer applications")

According to the search and conditions established, the following results were obtained. They are shown in Table 2.

Table 2. Studies per database

Studies per database	
Database	Search
IEEE	67 Relevant results
ACM – Digital Library	42 Relevant results
Scopus (ScienceDirect)	387 Relevant results
Inspec/Compendex	25 Relevant results

3.3 Study Selection and Quality Evaluation

In order to include the relevant works, selection criteria were applied to the titles and summaries. They present the following characteristics:

- The studies present the method and the research result
- The studies are in the field of citizen science
- The studies that were published online from 2017 (first systematic evaluation), then those made during 2016 and, finally, revision of citizen science in 2015.

The criteria for the exclusion of works were the following:

- Studies that present only summaries of conferences, editorials, templates
- Studies that do not present evidence of peer-review
- Studies not presented in English
- Studies not accessible in full text
- Unclear books and literature
- Duplicate studies of other related studies

4 Search Results

4.1 Frequency of Publication

The results obtained were 521, after that the method of inclusion and exclusion was applied, eliminating publications prior to 2015, reading full text, lightning sampling, qualitative evaluation, review of publications excluded, as shown in Fig. 1.

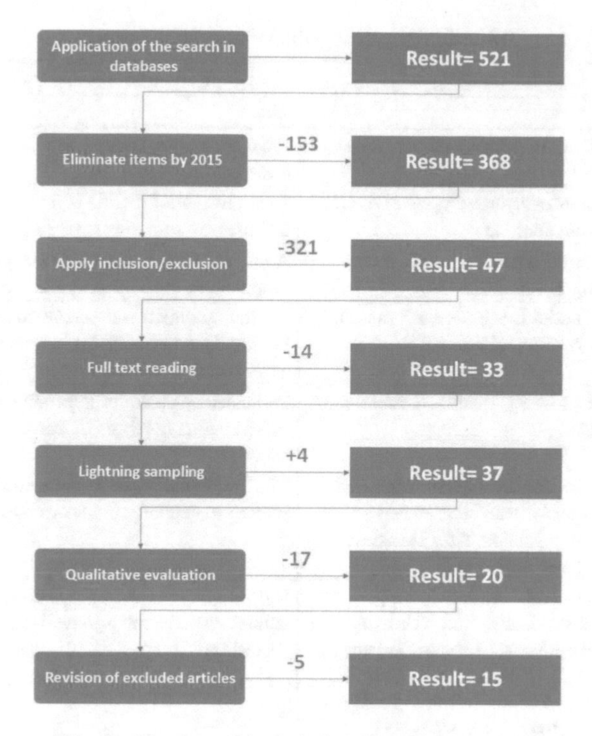

Fig. 1. Number of included and excluded articles

4.2 Extraction Information

Table 3 shows the studies selected based on the intervention of citizen science in the agricultural sector through the use of ICTs. According to the searches made in the different databases, a favorable result has been obtained in Scopus journals with more publications about citizen science in agriculture with the use of ICTs. The results are shown in Fig. 2.

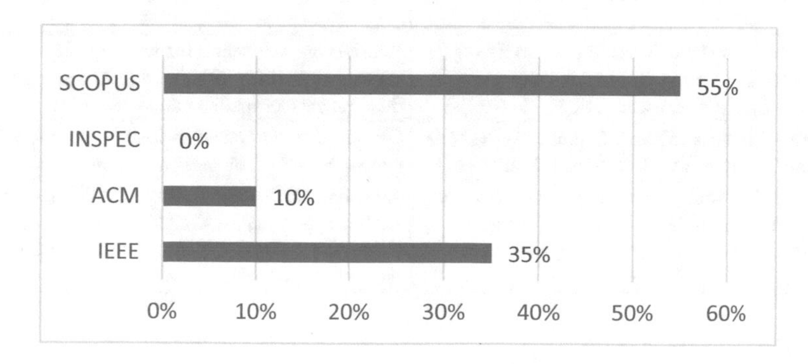

Fig. 2. Search results in databases

Table 3. Relevant publications for the search

Relevant results	
Author	Title
Martina Marjanović, Aleksandar Antonić, Ivana Podnar Žarko [6]	Edge computing architecture for mobile crowdsensing
Mahmudur Rahman, Mozhgan Azimpourkivi, Umut Topkara, Bogdan Carbunar [7]	Video liveness for Citizen journalism: attacks and defenses
Poonam Yadav, Ioannis Charalampidis, Jeremy Cohen, John Darlington, Francois Grey [8]	A collaborative Citizen science platform for real-time volunteer computing and games
Romani Luciana, Gabriel Magalhães, Martha D. Bambini, Silvio R. M. Evangelista [9]	Improving digital ecosystems for agriculture: users participation in the design of a mobile app for agrometeorological monitoring
Tian Cai, Hastings Chiwasa, Charles Steinfield, Susan Wyche [10]	Participatory video for nutrition training for farmers in Malawi: an analysis of knowledge gain and adoption
Steven Gray, Rebecca Jordan, Alycia Crall, Greg Newman, Cindy Hmelo-Silver, Joey Huang, Whitney Novak, David Mellor, Troy Frensley, Michelle Prysby, Alison Singer [11]	Combining participatory modelling and citizen science to support volunteer conservation action
Duncan C.McKinley, Abe J.Miller-Rushing, Heidi L.Ballard, Rick Bonney, Hutch Brown, Susan C.Cook-Patton, Daniel M.Evans, Rebecca A. French, Julia K.Parrish, Tina B.Phillips, Sean F. Ryan, Lea A.Shanley, Jennifer L.Shirk, Kristine F. Stepenuck, Jake F.Weltzin, AndreaWiggins, Owen D.Boyle, Russell D.Briggs, Michael A. Soukup [12]	Citizen science can improve conservation science, natural resource management, and environmental protection
Kieran Hyder, Serena Wright, Mark Kirby, Jan Brant [13]	The role of citizen science in monitoring small-scale pollution events
Matthias Schröter, Roland Kraemer, Martin Mantel, Nadja Kabisch, Susanne Hecker, Anett Richter, Veronika Neumeier, Aletta Bonn [14]	Citizen science for assessing ecosystem services: status, challenges and opportunities
Denis Couvet, Anne-Caroline Prevot [15]	Citizen-science programs: Towards transformative biodiversity governance
N.A.Welden, P.A.Wolseley, M.R.Ashmore [16]	Citizen science identifies the effects of nitrogen deposition, climate and tree species on epiphytic lichens across the UK
David G.Rossiter, Jing Liu, Steve Carlisle, A.-Xing Zhu [17]	Can citizen science assist digital soil mapping?
Ian Thornhill, Jonathan G. Hob, Yuchao Zhang, Huashou Li, Kin Chung Ho, Leticia Miguel-Chinchilla, Steven A.Loiselle [18]	Prioritising local action for water quality improvement using citizen science; a study across three major metropolitan areas of China
B. Weeser, J. Stenfert Kroese, S. R Jacobs, N. Njue, Z. Kemboi, A. Ran, M. C. Rufino, L.Breuer [19]	Citizen science pioneers in Kenya – a crowdsourced approach for hydrological monitoring
Mariette McCampbell, Marc Schut, Inge Van den Bergh, Boudy van Schagen, Bernard Vanlauwe, Guy Blomme, Svetlana Gaidashova, Emmanuel Njukwe, Cees Leeuwis [20]	Xanthomonas Wilt of Banana (BXW) in central Africa: opportunities, challenges, and pathways for citizen science and ICT-based control and prevention strategies

4.3 Results Analysis

In the preliminary database search of IEEE Xplore, ACM, Inspec / Compendex and Scopus (ScienceDirect), 521 preselected articles were obtained, of which 15 met the inclusion criteria considered in the present work. Among the results obtained, the articles can be categorized according to the application used, the citizen participation technique applied and the benefits generated. Hence, within the application 93% articles consider web portals, software use or their combination. Other identified technologies were motion sensors, remote sensors, online games, cloud computing and telecommunications equipment.

According to the category of citizen participation techniques and as expected in this study, group participation and collaborative work stand out as the most used techniques in 87% of the articles. Surveys and interviews, thematic workshops, volunteer work, experts and on-site data generation are other means of citizen participation that were considered.

There are multiple benefits generated by citizen participation, and they depend on the quality and quantity of the input. These include crop management, pollution control, flood control, environmental and meteorological monitoring, and the greatest amount (67%) of the benefits were related to learning systems, resource management and environmental conservation.

Table 4 synthesizes the information from citizen science studies applied to agriculture and its relation to the application of ICTs, citizen participation techniques and benefits generated. It can also be observed that the same article can make use of more than three ICT applications [6–9, 17], participation techniques [7, 9–11, 14, 15, 20] and more than three benefits [12, 13, 18]. From this it follows that the study conducted by Rahman et al. [7] and Romani et al. [9], have a greater number of ICT application criteria as well as citizen participation techniques, while the article by Thornhill et al. [18] presents a greater number of benefits, such as environmental and meteorological monitoring, flood control and learning systems.

5 Results Analysis: Discussion

Citizen science is becoming more popular thanks to its great potential in various applications. Based on the information provided by the community in general that goes through a process of validation and feedback to the user. It is important to add reward proposals to attract more collaborators. Hence, several scientific studies and their applications are based on citizen participation. In the field of agriculture, the applications of citizen science are wide. The selected articles list benefits such as environmental monitoring, pest control, water quality, monitoring of weather conditions, soil studies and, above all, the improvement of teaching systems.

Potential benefits could be considered timely identification of outbreaks of pests, shortage of water resources, pollution problems, efficient use of natural resources, agricultural insurance. According to the results obtained from the systematic review presented in this work, citizen science currently contributes to the environment and the technological solutions that are presented, facilitating their development. There are

Table 4. Analysis of relevant publications

Relevant publications		[6]	[7]	[8]	[9]	[10]	[11]	[12]	[13]	[14]	[15]	[16]	[17]	[18]	[19]	[20]
ICTs applications	Mobile devices	x		x	x										x	x
	Telecommunication equipments		x		x								x			
	Webs			x			x	x		x	X	x			x	x
	Cloud computing	x		x												
	Videos		x			x										
	Online games			x												
	Motion sensors		x													
	Remote sensors												x	x		
	Software	x	x		x		x		x			x	x	x		
Citizen participation techniques	Collaborative work	x	x	x		x	x		x	x	X	x	x	x	x	x
	Workshops for farmers				x			x								
	Experimental treatment		x		x	x										
	Volunteer work and formative education		x	x		x	x				X					
	Expert consulting				x		x	x			X					
	On-site data generation									x			x	x	x	x
	Surveys and interviews	x	x						x	x		x				x
Benefits	Environmental monitoring	x						x	x	x		x		x		
	Meteorological monitoring	x			x											
	Management of resources and environment preservation						x	x	x	x	X			x	x	
	Soil use							x					x			
	Pest control															x
	Crop management				x								x			x
	Pollution control											x		x		
	Flood control														x	
	Learning systems		x	x		x	x		x		X			x		

aspects that are related to citizen science, such as the participatory model that is often based on the structuring of environmental problems or scientific questions in a way that gives an answer to questions of scientific or social concern involving collaboration from both the public and professional scientists.

However, it must be taken into account that for the development of community projects, the ability to participate in data collection associations can improve the ability of participants to make decisions. The informatic tools have become a support for the new type of learning community that is being generated, among the most outstanding the web, mobile technology, wifi networks, sensors, modeling software and videos, which include the general public, and specialists in certain areas increasing skills and knowledge, therefore we must take into account that public participation can be sensitive when handling information that at a certain point can become fraudulent especially with the use of videos; On the other hand, the positive aspect is that it can become a mixed training method that integrates both the video training of a subject as well as the exchange of experience among professionals.

In studies such as the one conducted by Gray et al. [11], the different characteristics and strengths of both citizen science and participatory modeling are discussed and, despite handling similar objectives, there are approaches that integrate them into a single program. Therefore, they use both a modeling software and a web deploying the necessary tools to undertake an environmental management project integrated by a voluntary group that work with agricultural professionals, giving rise to the promotion of management plans. The analyses led to the application of case studies referring to the management of the habitat and to the quality of water and livestock. Likewise, an approach is presented that includes the participation of end users as representatives in the design process of a mobile application that supports farmers in the agro-meteorological monitoring of their crops [9]. This approach also uses aspects of Socionomy to improve user participation during technical meetings and evaluation sessions of the mobile application. The results of this research indicate that both users and developers benefit from this interaction once the product meet the needs, reducing the number of errors. The aim of the study conducted by Weeser et al. [19] was to involve citizens in a water level monitoring project in response to the scarcity of data in remote basins, such as the Sondu-Miriu river basin in Kenya. To collect and process observations, a software and hardware framework was developed based on an approach where text messages sent by the observers were used to transmit the collected data and installation of signs next to the water level indicators, explaining the system for interested participants. In this sense it can be applied in agricultural contexts that lack information for decision making. Citizen science programs are cost-effective methods to monitor environmental resources, which makes them especially suitable for low-income countries to overcome their limited data resolution. On the other hand, in the selection process of studies related to agriculture, articles focused on other applications were identified, which due to their methodology could be considered in future agricultural studies.

6 Conclusion and Future Research

This work shows a systematic review that allows the identification and evaluation of several citizen science studies related to agriculture, which were analyzed in terms of their results, and also in order to answer the questions initially posed. The results show that citizen science in conjunction with ICTs is a growing trend and that citizen knowledge can complement and improve the technological solutions of agriculture. In the reviewed studies it was observed that technology can solve problems that require specialized knowledge, in addition to helping the farmers in their decision making, thus avoiding waste of time and resources. The effect of citizen science on agriculture and ICTs is positive. Citizen science seeks to involve more farmers in research and development activities, creating opportunities for interaction and integration of the community in the improvement of crop fertility, soil management, monitoring of environmental resources, pest management, among others. Through citizen participation, new research problems arise that will allow finding solutions for the provision of services to farmers. In addition, ICTs allow the integration of different groups of specialized and non-specialized people who can contribute with new information, thereby generating a benefit that encompasses various sectors of the population, providing clear understanding and communication about agriculture.

As future work, it is possible to extend the search and include studies related to the environment and focused on specific areas, such as air, soil and water, as well as the incorporation of new digital libraries. Future studies will allow the development of new projects that combine citizen knowledge with technology and provide access and exchange of information, particularly, in rural areas where there are no specific studies either due to lack of resources or planning.

References

1. Finquelievich, S., Fischnaller, C.: Ciencia ciudadana en la Sociedad de la Información: nuevas tendencias a nivel mundial. *Revista iberoamericana de ciencia tecnología y sociedad,* **9**(27), 11–31 (2014)
2. Pérez, A., Milla, M., Mesa, M.: Impacto de las tecnologías de la información y la comunicación en la agricultura. *Cultivos Tropicales,* **27**(1), 11–17 (2006)
3. Nagel, J.: Principales barreras para la adopción de las TIC en la agricultura y en las áreas rurales. *CEPAL,* 54 (2012)
4. Kitchenham, B., Charters, S.: Guidelines for performing systematic literature. Technical report, EBSE Technical (2007)
5. Bazán-Vera, W., Bermeo-Almeida, O., Samaniego-Cobo, T., Alarcon-Salvatierra, A., Rodríguez-Méndez, A., Bazán-Vera, V.: The Current State and Effects of Agromatic: A Systematic Literature Review. Springer, Berlin (2017)
6. Marjanović, M., Antonić, A., Podnar Žarko, I.: Edge computing architecture for mobile crowdsensing. IEEE Xplore Digit. Libr. **6** (2018)
7. Rahman, M., Azimpourkivi, M., Topkara, U., Carbunar, B.: Video liveness for Citizen journalism: attacks and defenses. IEEE Xplore Digit. Libr. **16** (2017)

8. Yadav, P., Charalampidis, I., Cohen, J., Darlington, J., Grey, F.: A collaborative Citizen science platform for real-time volunteer computing and games. IEEE Xplore Digit. Libr. **5** (2018)
9. Romani, L., Magalhães, G., Bambini, M.D., Evangelista, S.R.M.: Improving digital ecosystems for agriculture: users participation in the design of a mobile app for agrometeorological monitoring. ACM Digit. Libr. 234–241 (2015)
10. Cai, T., Chiwasa, H., Steinfield, C., Wyche, S.: Participatory video for nutrition training for farmers in Malawi: an analysis of knowledge gain and adoption. ACM Digit. Libr. (2015)
11. Gray, S., et al.: Combining participatory modelling and citizen science to support volunteer conservation action. Biol. Conserv. ScienceDirect, **208** (2017)
12. McKinley, D.C., et al.: Citizen science can improve conservation science, natural resource management, and environmental protection. Biol. Conserv. ScienceDirect, **208** (2017)
13. Hyder, K., Wright, S., Kirby, M., Brant, J.: The role of citizen science in monitoring small-scale pollution events. Mar. Pollut. Bull. ScienceDirect, **120** (2017)
14. Schröter, M., et al.: Citizen science for assessing ecosystem services: status, challenges and opportunities. Ecosyst. Serv. ScienceDirect, **28** (2017)
15. Couvet, D., Prevot, A.-C.: Citizen-science programs: towards transformative biodiversity governance. Environ. Dev. ScienceDirect, **13** (2015)
16. Welden, N., Wolseley, P., Ashmore, M.: Citizen science identifies the effects of nitrogen deposition, climate and tree species on epiphytic lichens across the UK. Environ. Pollut. ScienceDirect, **232** (2018)
17. Rossiter, D.G., Liu, J., Carlisle, S., Zhu, A.-X.: Can citizen science assist digital soil mapping?. Geoderma ScienceDirect, **259–260** (2015)
18. Thornhill, I., et al.: Prioritising local action for water quality improvement using citizen science; a study across three major metropolitan areas of China. Sci. Total Environ. ScienceDirect, **584–585** (2017)
19. Weeser, B., et al.: Citizen science pioneers in Kenya – a crowdsourced approach for hydrological monitoring. Sci. Total Environ. ScienceDirect, **631–632** (2018)
20. McCampbell, M., et al.: Xanthomonas Wilt of Banana (BXW) in central Africa: opportunities, challenges, and pathways for citizen science and ICT-based control and prevention strategies. NJAS-Wagening. J. Life Sci. (2018)

Sentiment Analysis in Social Networks for Agricultural Pests

Oscar Bermeo-Almeida(✉), Javier del Cioppo-Morstadt, Mario Cardenas-Rodriguez, Roberto Cabezas-Cabezas, and William Bazán-Vera

Faculty of Agricultural Sciences, Agrarian University of Ecuador, Av. 25 de Julio y Pio Jaramillo, P.O. BOX 09-04-100, Guayaquil, Ecuador
{obermeo,jdelcioppo,mcardenas,rcabezas,wbazan}@uagraria.edu.ec

Abstract. Nowadays a large amount of subjective information is generated through social networks such as Facebook® and Twitter®. However, analyze manually this information require effort and time. Therefore, multiple sentiment analysis approaches are being proposed as a solution to this problem. Sentiment analysis is the field that studies opinions, feelings, sentiments, and attitudes that people express towards different topics of interest. Agriculture domain implies a large area of opportunity to obtain benefits using sentiment analysis, such as obtaining information about insects that affect sugarcane, rice, soya, and cacao crops, chemical substances used in crop diseases control and management, symptoms, recommendations, treatment, among others. However, agriculture domain has been very little studied. In this sense, we propose a sentiment analysis approach for agriculture to obtain the polarity at the comment and entity levels from texts. Finally, we assess the performance of our system under precision, recall, and F-measure metrics, obtaining average values of 77.43%, 77.50% and 77.35%, respectively.

Keywords: Agriculture · Agricultural pests · Sentiment analysis Social networks

1 Introduction

Nowadays, a huge amount of subjective information is generated through social networks such as Twitter® and Facebook®. The subjective information implies the opinions, beliefs, and attitudes that people express towards different topics of interest and it has a radical importance for different companies, organizations or individuals because it gives them knowledge about what people express towards the products or services they offer, which enables them to carry out actions that generate benefits for them, such as how to make better decisions, improve advertising campaigns or business strategies, among others.

R. Valencia-García et al. (Eds.): CITAMA 2019, AISC 901, pp. 122–129, 2019.
https://doi.org/10.1007/978-3-030-10728-4_13

Sentiment analysis or opinion mining is the field that studied opinions, sentiments and attitudes that people express towards different topics of interest. There are different sentiment analysis approaches namely lexicon based and machine learning; and three sentiment analysis levels: document, sentence, and entity.

On the other hand, sentiment analysis is used in a great diversity of domains, such as tourism, politics, health, among others [4–6]. However, the agriculture domain has been little explored although it has many benefits to offer, for example, through sentiment analysis in that domain it is possible to know the farmers and another people opinions about insects that affect sugarcane, rice, soya, and cacao crops, chemical substances used in crop diseases control and management, symptoms that crops present when are affected by an insect pest, among others. With this information, it is possible to take actions to avoid using chemicals that cause harm to health, take actions to prevent pests in crops, know what is the best treatments for crop diseases caused by an insect pest, to mention a few examples.

Based on this knowledge, this paper proposes a sentiment analysis approach that analyze English tweets and post related to agriculture domain, specifically, about pests control in crops, aiming that different farmers and people interesting in agriculture benefit from the information obtained through sentiment analysis. Also, it is important to mentioned that this work is based on English language due to the amount of information generated in this language day by day.

This document is structured as follows. Section 2 discusses the related works on the use of sentiment analysis. Section 3 describes the design of our approach for polarity identification, Sect. 4 presents evaluation and results and Sect. 5 presents the conclusions and future work.

2 Related Word

Research in sentiment analysis started in the early 2000s. Since then, several methods to analyze the opinions and detect polarity from online opinion sources such as blogs, forums, tweets and Facebook post have been introduced. Two main approaches have been introduced: Semantic orientation approach and machine learning approach. In the first approach, commonly a sentiment lexicon is used such as SentiWordNet [7, 11], iSOL, eSOL [6], and Ml-Senticon [10]. SentiwordNet is the most popular lexicon used in the literature. The process of this approach consists in search each word in the lexicon and assign a positive, negative or neutral value and obtain an average that represent the global polarity of the text. On the other hand, in the machine learning approach supervised classification algorithms such as Support Vector Machine (SVM) [2, 8, 13], Naive Bayes (NB) [7, 18], BayesNet (BN) [12], and Maximum Entropy (MaxEnt) [4] are used. The classification algorithm needs a training set to build a model from diverse features of the corpus documents and a testing set to validate the built model from the training set.

With regards sentiment analysis levels, most works have focused on document-level polarity identification [1], where a polarity positive, negative, or neutral is assigned to the whole document, whilst, few approaches have based their studies in a sentence-level and entity-level.

Finally, regarding domain most of proposals are based on context such as movies, hotels, health, and products. However, in agriculture domain only one work has been presented in the last years [17]. In this work, authors propose a framework for applying opinion mining in blogs for agriculture. Data mining techniques and sentiment analysis are applied to extract text from blog concerning agriculture to know the opinions about agriculture subjects.

In this paper, a comment-level and entity-level sentiment analysis are addressed. Also, we propose an approach for agriculture domain which has been considered by very few works. We consider that entity-level and comment-level sentiment analysis in agriculture domain is of utmost important since a tweet or post could contain positive, negative or neutral opinions about several aspects in the text.

3 Approach

The proposed sentiment analysis approach is divided into three main components: (1) data collection, (2) preprocessing module, and (3) polarity identification. Figure 1 shows the architecture of the system. The first module consists in collect data sets

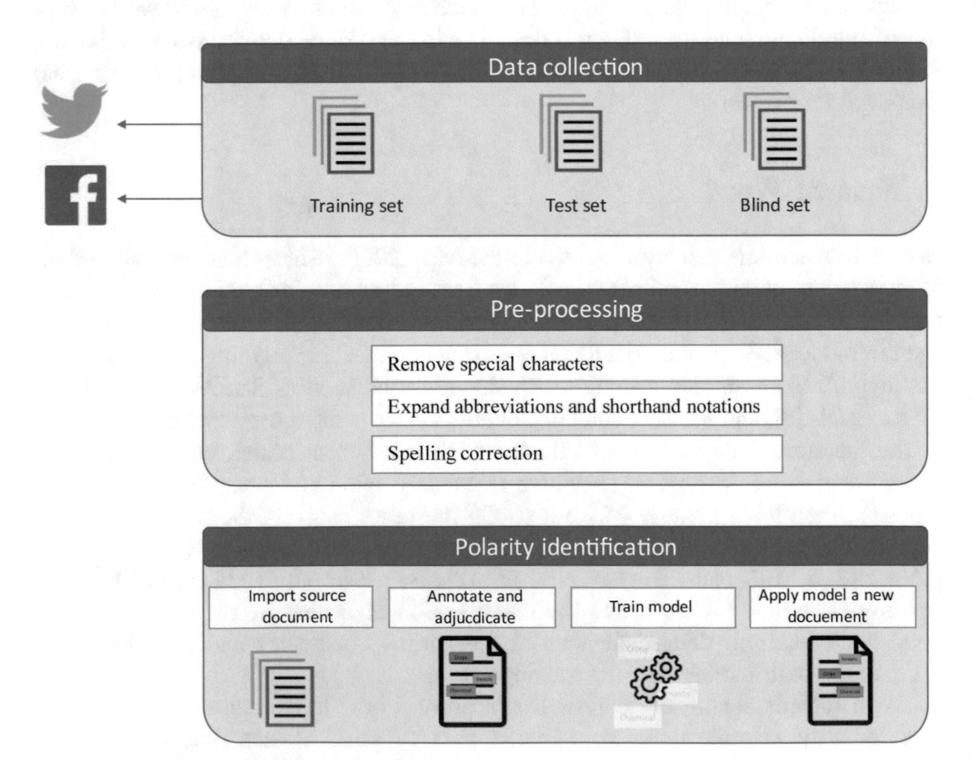

Fig. 1. Our approach.

social networks such as Twitter and Facebook. The second module involves the pre-processing of the corpus to clean and correct the text. The last module calculates the polarity (positive, negative, or neutral) of each post o tweet. A detailed description of the modules contained in the architecture is provided in the following sections.

3.1 Data Collection

Since this work is based on Machine learning approach, we require a set of data to training an algorithm. We use Facebook Graph API [3] and Twitter API [16] to respectively extract information from Facebook and Twitter about (1) insects that affect sugarcane, rice, soya, and cacao crops; (2) chemical substances used in crop management; (3) diseases, symptoms, treatments and recommendations for crops control and management.

The corpus consists of a total of 300 tweets and Facebook post, which were manually tagged by a group of three two in agriculture. The process involves reading the tweets and identifying all aspects and associated polarities (positive, negative, or neutral).

Next, Fig. 2 shows an example of a tweet about chemical used in cultives.

Fig. 2. Example of a tweet.

As can be seen in Fig. 1, our approach requires of three data sets: training set, test set and blind set. The first one consists of a set of documents that have been annotated through expert's annotators in agriculture that is used to train the annotator component. The goal of the training set is to teach the machine-learning model about correct annotations. The second one consists in a set of annotated documents that is used to test the trained annotator component. Close analysis helps to find weaknesses in the current model that can be addressed. Finally, the third set consists of a set of annotated documents that is set aside and used to test the system periodically after several iterations of testing and improvement have occurred. To prevent accuracy from being tainted (for example, by making changes based only on annotations in known documents), blind data should be data that has not previously been viewed by users involved with creating the annotator component.

3.2 Pre-processing Module

Social network messages such as tweets and Facebook post vary in content and composition, often containing non-standard words, ungrammatical sentence structures and domain-specific terms, also known as jargon. This phenomenon reduces the accuracy in many natural language processing tasks. Aiming to address this issue, the present module focuses on normalize Twitter messages, i.e., normalize non-standard words to their canonical form. Furthermore, it removes characters that do not contribute to the meaning of the tweet. More specifically, this module performs three tasks:

1. Remove special characters that do not provide relevant information for the opinion mining and semantic profiling module. In this step strings such as URLs are removed.
2. Expand abbreviations and shorthand notations by their expansions. Abbreviations and acronyms are commonly used on the Web.
3. Spelling correction. We used the spell checker and morphological analyzer Hunspell [9] to correct spelling errors.

3.3 Polarity Identification

The polarity identification module was implemented by using the Natural Language Understanding® [5]. The sentiment analysis is carried out at the comment level and at the entity level, that is, the obtaining of sentiments (sentimental polarity), and domain-specific entities. It is important to mentioned that the polarity is obtained on a scale from −1 to 1, where a value equal to 0 indicates neutral polarity, a value greater than 0 indicates positive polarity and a value less than 0 indicates negative polarity.

Natural Language Understanding provides easy-to-use tools for annotating unstructured domain literature, and uses those annotations to create a custom machine-learning model that understands the language of the domain. The accuracy of the model improves through iterative testing, ultimately resulting in an algorithm that can learn from the patterns that it sees and recognize those patterns in large collections of new documents. In this sense, we develop our own model from corpus presented in Sect. 3.1.

4 Evaluation and Results

The experiments conducted to test our approach involved the use of a test set about inset pest in crops. Due to a lack of available corpora and datasets in English for this domain, we provided the platform with our own collection of texts (see Sect. 3.1 Data collection). The test set generation process is the following:

1. For three months, we collected a set of tweets and post.
2. Duplicate comments were removed by means of an automatic filter.
3. Two experts in the agricultural domain analyzed the comments and discarded those that did not denote an opinion.
4. Every comment was manually reviewed and labeled by the experts in terms of its polarity: positive, negative, or neutral. In total, 300 social network texts were collected: 100 positive, 100 negative, and 100 neutral.
5. We use inter-annotator agreement measure to ensure consistent annotations. The agreement calculated at this stage using the Cohen's κ score was satisfactory with a $\kappa = 0.65$.

To assess the performance of our system, the precision, recall, and the F-measure metrics were used. These metrics were proposed by [15] and are commonly employed to validate text classification systems, including opinion mining systems. Precision represents the proportion of predicted positive cases that are real positives (see Eq. 1), whereas recall is the proportion of actual positive cases that were correctly predicted as such (see Eq. 2). Finally, the F1 is the harmonic mean of precision and recall (see Eq. 3).

$$\boldsymbol{Precision} = \frac{\boldsymbol{True\,positives}}{\boldsymbol{True\,positives} + \boldsymbol{False\,positives}} \qquad \text{(Eq.1)}$$

$$\boldsymbol{Recall} = \frac{\boldsymbol{True\,positives}}{\boldsymbol{True\,positives} + \boldsymbol{False\,positives}} \qquad \text{(Eq.2)}$$

$$\boldsymbol{F1} = 2 * \frac{\boldsymbol{Precision} * \boldsymbol{Recall}}{\boldsymbol{Precision} + \boldsymbol{Recall}} \qquad \text{(Eq.3)}$$

In a multiclass classification, precision, recall, and the F-measure are calculated for each class (i.e. positive, negative, and neutral). Therefore, to generate an overall evaluation of our system, the evaluation results from each class were combined. To this end, we applied the macroavering metric (Lewis 1992), which is the arithmetic mean of precision, recall, and the F1, where the quotient is the number of classes used in the prediction. In this sense, the Macro-Precision and Macro-Recall equations can be proposed as follows:

$$\boldsymbol{Macro} - \boldsymbol{Precision} = \frac{\sum_{i=1}^{|C|} \boldsymbol{Precision}}{|\boldsymbol{C}|} \qquad \text{(Eq.4)}$$

$$\boldsymbol{Macro - Recall} = \frac{\sum_{i=1}^{|C|} \boldsymbol{Recall}}{|\boldsymbol{C}|} \tag{Eq.5}$$

The macro average F1 is the harmonic mean of the macro-precision and macro-recall scores.

Table 1 below summarizes the results at comment level for precision, recall, and the F-measure. The first column represents the three classes considered, whereas the following columns list the scores obtained for each class, as well as the system's overall performance score.

Table 1. Evaluation results

	Precision	Recall	F-measure
Positive	0.7557	0.8	0.7773
Negative	0.8071	0.85	0.828
Neutral	0.7601	0.675	0.7152
Macroavering	**0.7743**	**0.775**	**0.7735**

As can be seen in Table 1, the system's average scores for precision, recall, and the F1 are 77.43%, 77.50%, and 77.35%, respectively. Such results are encouraging, as they demonstrate that the system can successfully detect the polarity of tweets and post. Specifically, the system is able to detect tweets and negative post in a better way than positive and neutral ones. Finally, the neutral class had the least favorable results among the three classes, thereby implying that neutral opinions are a challenge for our sentiment analysis system. In fact, it is usually difficult to distinguish between neutral sentiment and non-sentiment bearing sentences.

5 Conclusions

In this paper, we present a sentiment analysis approach for agriculture domain. Our approach relies on machine-learning-based opinion mining techniques to identify the polarity (positive, negative, or neutral) from Tweets and posts of Facebook. It also allows obtaining domain-specific entities in addition to the percentage of polarity for each one of them. In this sense, the system allows farmers to know other people opinions about insects that affect sugarcane, rice, soya, and cacao crops, chemical substances used in crop diseases control and management, symptoms that crops present when are affected by an insect pest, among others. To assess the performance of our system, we conducted a set of experiments on a corpus of 300 tweets and post (100 positive, 100 negative, and 100 neutral). Our approach obtained encouraging performance results, with average scores of 77.43% for precision, 77.50% for recall, and 77.35% for the F-measure. Similarly, our findings suggest that the classification of neutral-sentiment bearing content is still a challenge for our approach. As a future work, we have planned to apply this method to other languages such as Spanish and

French. Also, it is considered the identification of other important aspects in the text, such as irony, sarcasm, satire, among others.

References

1. Biyani, P., et al.: Co-training over domain-independent and domain-dependent features for sentiment analysis of an online cancer support community. In: Proceedings of the 2013 IEEE/ACM International Conference on Advances in Social Networks Analysis and Mining – ASONAM 2013, pp. 413–417. ACM Press, New York, USA (2013)
2. Cruz, N.P., et al.: A machine-learning approach to negation and speculation detection for sentiment analysis. J. Assoc. Inf. Sci. Technol. **67**(9), 2118–2136 (2016)
3. Facebook: API Graph
4. He, Y., Zhoub, D.: Self-training from labeled features for sentiment analysis. Inf. Process. Manag. **47**(4), 606–616 (2011)
5. IBM: Natural Language Understanding
6. Molina-González, M.D., et al.: Semantic orientation for polarity classification in Spanish reviews. Expert Syst. Appl. **40**(18), 7250–7257 (2013)
7. Montejo-Ráez, A., et al.: Ranked WordNet graph for sentiment polarity classification in Twitter. Comput. Speech Lang. **28**(1), 93–107 (2014)
8. Moraes, R., et al.: Document-level sentiment classification: an empirical comparison between SVM and ANN. Expert Syst. Appl. **40**(2), 621–633 (2013)
9. Németh, L.: Hunspell
10. Ofek, N., et al.: Improving sentiment analysis in an online cancer survivor community using dynamic sentiment lexicon. In: 2013 International Conference on Social Intelligence and Technology, pp. 109–113. IEEE (2013)
11. Peñalver-Martinez, I., et al.: Feature-based opinion mining through ontologies. Expert Syst. Appl. **41**(13), 5995–6008 (2014)
12. del Pilar Salas-Zarate, M., et al.: A study on LIWC categories for opinion mining in Spanish reviews. J. Inf. Sci. **40**(6), 749–760 (2014)
13. Rushdi Saleh, M., et al.: Experiments with SVM to classify opinions in different domains. Expert Syst. Appl. **38**(12), 14799–14804 (2011)
14. Sabra, S., et al.: Prediction of venous thromboembolism using semantic and sentiment analyses of clinical narratives. Comput. Biol. Med. **94**, 1–10 (2018)
15. Salton, G., McGill, M.J.: Introduction to Modern Information Retrieval. McGraw-Hill, Maidenheach (1983)
16. Twitter: Twitter API
17. Valsamidis, S., et al.: A framework for opinion mining in blogs for agriculture. Procedia Technol. **8**, 264–274 (2013)
18. Xia, R., et al.: Ensemble of feature sets and classification algorithms for sentiment classification. Inf. Sci. (NY). **181**(6), 1138–1152 (2011)

Trends in the Use of Webapps in Agriculture: A Systematic Review

Mariuxi Tejada-Castro[1(✉)], Carlota Delgado-Vera[1], Mayra Garzón-Goya[1], Andrea Sinche-Guzmam[1], and Xavier Cárdenas-Rosales[2]

[1] Escuela de Ingeniería en Computación e Informática, Facultad de Ciencias Agrarias, Universidad Agraria del Ecuador, Av. 25 de Julio y Pio Jaramillo, 09-04-100, Guayaquil, Ecuador
{mtejada, cdelgado, mgarzon, asinche}@uagraria.edu.ec
[2] Agrosoft, Urdesa Central, Circunvalaciòn Sur 1004 e Ilanes, Guayaquil, Ecuador
xcardenas@agrosoft.com

Abstract. Currently, the use of technology in agriculture is increasing, playing a fundamental role. Its use is becoming more and more widespread as farmers' demands grow. The employment of WebApp is increasingly transforming the way in which information is disseminated and obtained in the agricultural sector. In this sense, this work presents a systematic review of the literature on the WebApp's tendency in agriculture. Its objective was to identify in which phases of the crop cycle it has the most technological support, Web or mobile, and what functionalities the applications carry out, as well as to detect the tendency of use by the farmer. Accordingly, tools were used that allow us to make descriptive statistical metrics, where they proved that, due to their versatility and multi-platform, the web applications are fulfilling this objective, covering in its entirety all the phases of the crop. The countries that most often use them are Spain, Mexico, Colombia and the US.

Keywords: Phases of the crop · Web applications · Mobile applications Mobile and web technologies

1 Introduction

Agriculture has accompanied the man for thousands of years, and its processes have evolved over time. According to the FAO, a third of the global population derives its livelihood from agriculture, and in emerging economies it can represent up to 30% of GDP [1]. The National Development Plan states that innovation should offer the possibility of applying new productive techniques that include the rescue and validity of ancestral practices, as well as innovations. Institutions that make it feasible the transformations required in the Peasant Family Agriculture and subsistence agricultural systems in general. Therefore, this sector requires the integration of different actors to achieve modernization. In recent years, private and public companies belonging to the industrial, agricultural and ICT sectors have joined forces [2].

R. Valencia-García et al. (Eds.): CITAMA 2019, AISC 901, pp. 130–142, 2019.
https://doi.org/10.1007/978-3-030-10728-4_14

Software tools have become a fundamental complement in the monitoring and control of agricultural crops [3], becoming a great support for decision-making in the interest of improving productivity. Their information system has a database which suggests the most convenient path to follow and although the importance in the use of these tools is recognized, many farmers and peasants show a low interest in its use [4].

Later, they began to use ICT to improve soil and crop management; as a result, the applications were extended and adapted to different crops, products and countries. Precision Agriculture [5] is defined as a set of procedures and processes that seek to spatially and temporally optimize the life cycle of different crops through technologies, elements and studies in an environmentally friendly manner.

Nowadays the use of mobile applications is a technical as well as a commercial necessity in the different áreas. The article "Mobile Web Apps" indicates that there are different platforms and criteria that help us to reduce our options to select the most suitable development approach for a given situation [6]. The use of mobile applications allows farmers to detect diseases in crops, using an image captured by the cell phone in real time. With the techniques used in the image processing as a histogram and the color information of the captured image, vegetation indices are obtained which use image processing algorithms, for which high quality images are required.

There are both web and mobile applications that do not require inter-net connectivity to process the image inside the cell phone and serve farmers in the absence of Internet connectivity.

2 Analysis and Planification of the Review

In order to establish a starting point around the knowledge generated in the field of information systems based on precision Agriculture technologies, Kitchenham's proposal for the preparation of the literature review was followed [7]. Although this proposal is focused on revisions in software engineering, it is adaptable to other topics. The procedure to carry out the review consists of five steps: defining research questions, carrying out the literary search, selecting studies, classifying articles, extracting and carrying out the aggregation of data.

2.1 Questions

In order to carry out this research, three guide questions were made:

RQ1: What webapp tools are used / available in the context of agriculture?

RQ2: What functions do the software tools perform according to each phase / cycle in the context of agriculture?

RQ3: What is the trend of use of webapps applications?

2.2 Sources

Table 1 shows the sources where the systematic literature review was carried out. They include virtual libraries that are listed below. It specifies the type of bibliographic source and the period of publications of the archives. It is necessary to mention that the review was made in articles that contain the keywords and according to the search strategies established.

Table 1 Search and source chain

Virtual Libraries	Type of Bibliographical source	Search applied to	Language
Science Direct	Electronic books	Key words	English
IEEE Explorer	Scientific papers		
Elsevier	Conferences		
Springer			
ACM DL Digital Library			

2.3 Search Strategies

In order to answer the research questions, search criteria such as strategies were applied in the most relevant research works between 2013 and 2018. The search chain used is as follows:

- (crop cycle) and (functions of mobile and web applications)
- (Technology in precision agriculture) or (web) or (app móvil)
- (trends of web applications) and (app in precision agriculture)

2.4 Exclusion Criteria

In this systematic review, we are interested only in the articles that present the use as well as the development of applications that help in some of the processes of the crop cycle. Applications can be commercial, free and/or research. Therefore, we browse Google Scholar and SCOPUS Database using the keywords "agriculture" and variations of "application (s).

3 Development and Literature Review

This section presents the execution of the systematic review that consisted in the search of research works related to trends in the use of webapps applications in selected digital libraries, considering the inclusion and exclusion of criteria. In addition, this review allowed responding to the research questions presented in Sect. 2.1. These answers are discussed in the following section.

3.1 RQ1: What Webapp Tools Are Used/Are Available in the Context of Agriculture?

Once the webapps applications were mapped, Fig. 1 shows that 55% of the 178 primary studies with respect to the WebApps provide assistance in a specific area (1), and 45% integrate or provide solutions for different functions in the crop cycle (2). These results indicate that there are only a few applications that provide a single function with respect to the complete crop cycle. This means that the WebApps creation trend is focused on multiple areas of the crop cycle, they are more versatile and allow many solutions to be offered in the field and in the administrative area.

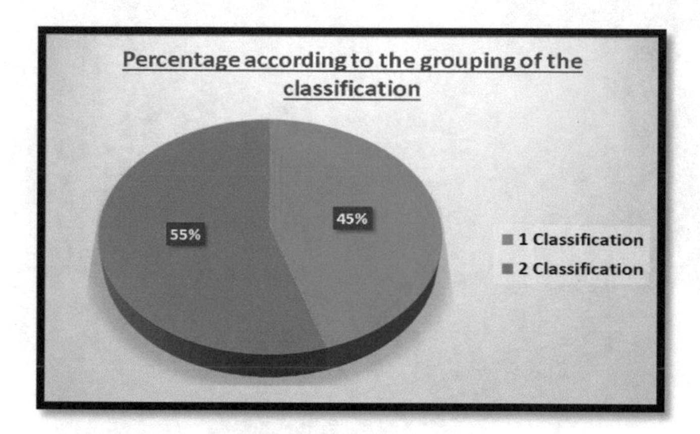

Fig. 1. Tool classification according to their functions (1) just one (2) several ones

3.2 RQ2. What Functions Do the Software Tools Perform According to Each Phase/Cycle in the Context of Agriculture?

In this section, as shown in Fig. 2, web and mobile applications are included according to the main activities involved in planting an agricultural product, such as: soil preparation, planting, fertilization, weed control, pests and diseases, irrigation, harvest, storage and sale [8]. Works based on precision agriculture technologies and procedures that have resulted from the search strategy and research questions were considered.

Applications for soil preparation. These types of applications provide global predictions for the standard numerical properties of soil (organic carbon, bulk density, cation exchange capacity (CEC), pH, soil texture fractions and coarse fragments) at seven depths, in addition to the predictions of depth to the bedrock and the distribution of soil classes based on the classification systems of the World Reference Base (WRB) and the USDA. As for the web applications, they allow to satisfy the diverse ways in which the users interact with the soil survey data, mainly informing how soil map units are designed and imitating how a soil survey is used.

Applications for the calendar of planting and establishment of the soil. There are systems developed for the management of the most efficient agricultural sowing processes thanks to the inclusion of mobile technologies in the field. They allow to get real data from the crops through web and mobile platforms designed under web services [9]. These applications offer services to the agricultural industry that serve as support in the management of the seed process in genetic resource banks; thus they contribute to deepen into the knowledge of mobile computing oriented to the support of agricultural

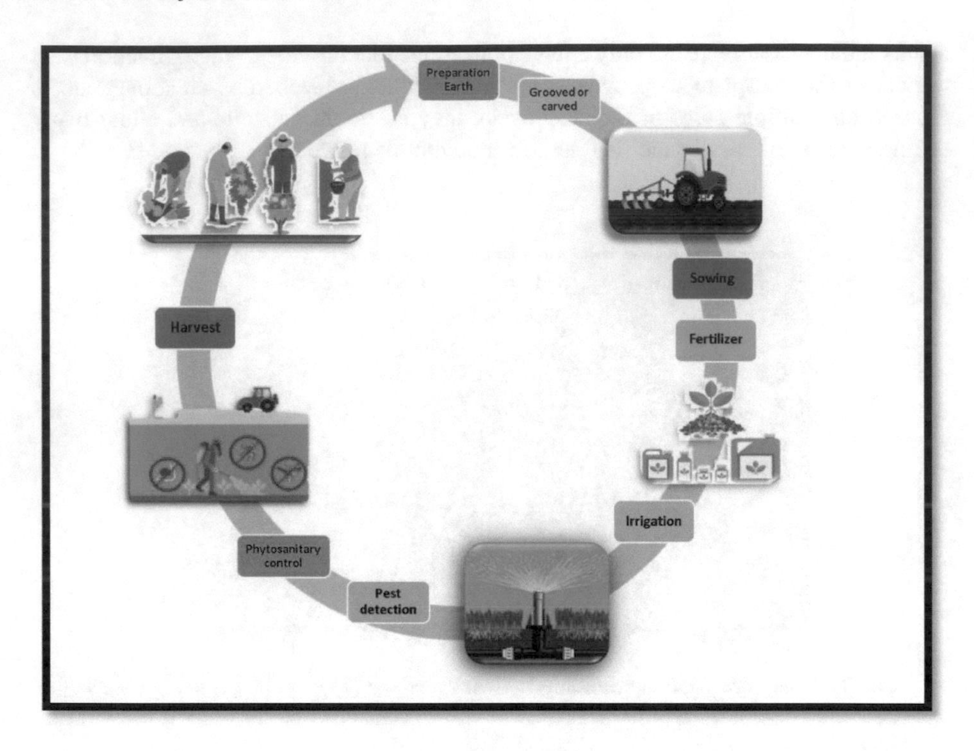

Fig. 2. General scheme of the crop process

processes in the field and serve as a model for the implementation of other applications in this sector [10].

Applications for crop fertilization and nutrition. Calculating the amount of fertilizers needed is another solution provided by the technology [11], which allows obtaining better combinations of fertilizers for the solution of nutrients in different crops. The current price of fertilizers in the market is taken into account, providing farmers and agricultural technicians with a tool to support agricultural tasks in situ. Web applications, databases and advanced mobile systems facilitate real-time data collection for effective monitoring; In addition, open source systems save money and facilitate a greater degree of integration and better development of applications based on the robustness of the system and its wide utility for various fields of engineering [5].

Applications for phytosanitary control of crops. In this process you can find applications that through the recognition of images help in the detection of pests, diseases, weeds, and damages. They also help with the methods and management of control of these diseases. The data are recorded, stored, processed and compared offering a diagnosis and / or solution to the damages caused by pests or diseases of the crop. There are applications that have been created by companies, others by universities and by research centers.

Applications for irrigation and drainage. In this category, the applications carry out several actions: topographic survey of the plots, registration of activities, measurement of perimeters, obtained yields, geo-positioning of soils, elaboration of maps. All this is done thanks to the arrival of advanced technology applications such as integrated computation, robotics, wireless technology, GPS/GIS (geographic positioning system/geographic information system), which are considered recent developments [12, 13].

Applications for management. Since the origin of agriculture, this has depended on the human hand and the limited use of equipment and machinery. The administration/control of their farms was almost nil. For this reason, there are web and mobile applications that support the farmer in these tasks, thus controlling their products, salaries, production costs, and traceability, becoming fundamental for farmers. These systems allow the capture, analysis, storage, communication of data, and the integration of supply chains which are essential for production success since the purpose is to make a complete follow-up of the entire life cycle of the product [14].

Regarding the topics presented, Table 2 shows the percentage of webapps applications that exist in the market grouped according to the phases of the crops, being the most used to support farmers in the planting, production, pests phase, irrigation, production and harvest.

Table 2 Distribution of WebApp applications according to the phases of the crop

Classification	WebApp	%
Fertilization	[15, 16, 17, 18, 19, 20, 21]	17%
Pests	[22, 23, 24, 25, 26, 27, 28, 29]	19%
Production and Harvest	[30, 31]	16%
Irrigation, Fertilization and Sowing	[32, 33, 34, 35, 36, 37]	22%
Sowing and production	[38, 39, 40, 41]	21%
Ground	[42, 43, 44]	5%

Table 3 shows the platform that WebApps use to control the different stages of agricultural crop production: Android, IOS and Web. It can be seen that 12% represent those that were developed in the Android operating system, 3% represent those developed in IOS, and 14% for both platforms. The web platform covers the highest percentage with 71%.

Table 3 Types of platforms

Platform	%
Android	12
Android/IOS	14
IOS	3
Web	71

3.3 RQ3. What is the Trend of Use of Webapps Applications?

The Google Trends tool was used (Fig. 3) to analyze the search trend of the keywords App crops and Web crops on the Internet during the period from January 1 2017 to August 31 2018. It was observed that Web applications for crops have a greater tendency in searching compared to mobile applications. Therefore, it can be considered that there is a greater demand for the use of applications in the web environment because in this environment it is possible to work with more integral processes and with the administration and control of all the processes involved in an agricultural crop, unlike mobile applications that are designed mostly for specific functions based on the stages of a crop.

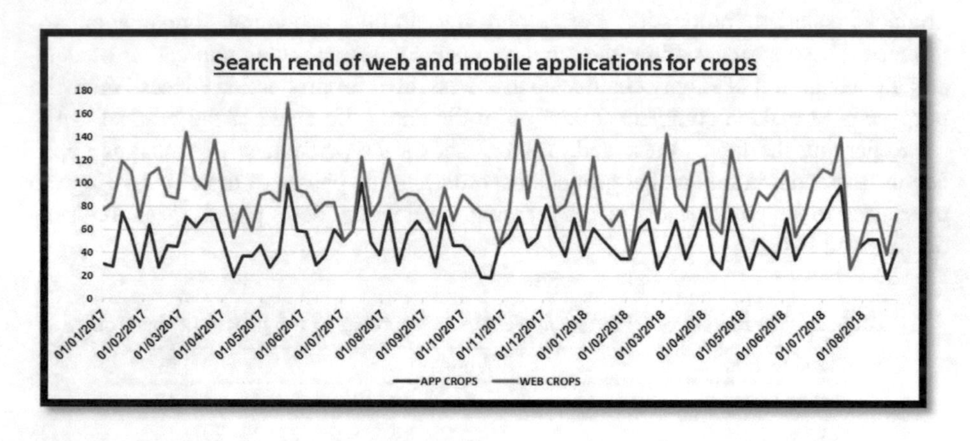

Fig. 3. Search trend for WebApp applications worldwide

4 Results

As seen in Table 4, these metrics establish the trend of use of mobile applications according to the score. They were grouped according to their field of application, that is, the positioning; the Sensor-Tower tool was used, which was created by a private company based in San Francisco and was a member of the AngelPad startup incubator program in 2013 [45].

Table 4 Trend of use of Apps according to the field of application

Phase	Amount App	Average	Min	Max	Median
Fertilization	7	22,14	1	42	21
Pests	4	43,00	37,00	56,00	39,5
Production and Harvest	2	20,50	12,00	29,00	20,5
Irrigation	2	12,50	8,00	17,00	12,5
Irrigation, Fertilization and Sowing	2	38,50	24,00	53,00	38,5
Sowing and production	3	39,33	24,00	49,00	45
Soil	2	69,50	61,00	78,00	69,5

The metrics for each of the web applications are analyzed. The tools SimilarWeb [46] and Alexa were used, which provided information on the metrics to be included in this analysis. Aspects of the web application were considered as the traffic ranking worldwide [47, 48], by category and country, as well as percentages of web search, averages of visits and the rebound indicator. The total web applications analyzed were 114, classified by each of the phases. It was observed that most correspond to the pest and crop phases, concentrating between them approximately 61% of the analyzed applications. On the other hand, Table 5 shows the frequency of web applications that have the Bounce Rate indicator. Only 39% of these applications of the study have this measure.

The explanation and consideration of each of the Metrics in Table 6 are detailed below:

GLOBAL TRAFFIC RANK: Site traffic rank, compared to all other sites in the world.

COUNTRY RANK – NAME: Name of the leading country of the visitor of the site.

COUNTRY RANK: Site traffic rank, compared in the leading country.

TRAFFIC SOURCES SEARCH: It's the percentage of site visitors that used search engines to find this site.

CATEGORY RANK: Site traffic rank, compared to all other sites in its main category.

HEADQUARTER: Name of the host country of the web application.

AVERAGE VISIT DURATION (minutes).

Period of time during which a user interacts with a page. Daily Time on Site: Estimated daily time on site (hh:mm:ss) per visitor to the site

BOUNCE RATE: The percentage of sessions on the page, that is, sessions in which the user has left the website on the login page without interacting with it. It indicates if the page is interesting for users from the first visit. If the bounce rate is very high, it means that many users are fleeing the page as they arrive. When analyzing the metrics and their descriptive statistics for each of the groups of web applications it was obtained that globally the web applications have an average GLOBAL TRAFFIC RANK indicator of 8,384,035 with a minimum value of 164. This indicates the position of the application AGROCON based in the UNITED STATES and with a greater use of this application at INDIA level. It corresponds to the group of applications analyzed of the planting phase with the best position in this ranking, while the maximum value obtained in this ranking indicates the lowest position and is associated with the AQUARIEGO application whose country with the highest use is registered in Peru.

Regarding COUNTRY RANK - NAME, the leading countries of use were identified by application groups. Web applications for which no information could be obtained have been discarded, so only 78 applications are considered. It is observed that the country with the greatest leadership in the use of web applications is Spain, concentrating 23% of the applications studied. Mexico follows with 15% web applications. Colombia and the United States are also registering leadership in the use of study applications. In general, 22 countries appear with participation in the use of these web applications.

COUNTRY RANK is related to the previous list, from which it can be mentioned that the minimum of this indicator corresponds to the position 41 of the ranking at the

Table 5 Metrics used in the Web pages

PHASE	FREC ABS	FREC REL	FRE REL_ PORC	FREC REL_PORC_ACU	WITH METRICA BOUNCE RATE DATA	NO METRICA DE BOUNCE RATE DATA
SOWING	20	0.175438596	17.54%	17.54%	12	8
IRRIGATION	25	0.219298246	21.93%	39.47%	11	14
PESTS	37	0.324561404	32.46%	71.93%	12	25
CROP	32	0.280701754	28.07%	100.00%	10	22
TOTAL	114	1	100%		45	69

country level of the application. This value is associated with AGROCON, a web application that also has the highest position of the studied ones. According to the world ranking, the maximum value for this indicator is 4,196,649 and corresponds to the position in the ranking according to the country that leads its use of the IRRICAD application, whose leading country is the United States.

Figure 4 also shows the TRAFFIC SOURCE SEARCH indicator, obtaining that the analyzed web applications have access from 74% of search engines. The histogram presents the distribution of the data of this metric where it is visualized that the concentration of the majority of web applications is about 60%.

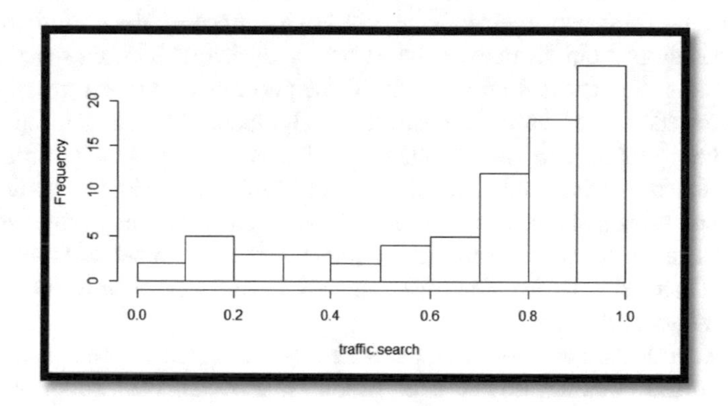

Fig. 4. Histogram of traffic source search

Regarding CATEGORY RANK, there is information on 68 applications, of which the application that has the best position in its category is INFOAGRO, based in Spain. The country that leads its use is Mexico. The application with a lower position in the ranking of categories is CPA PEST CONTROL AND ENVIRONMENT, about which there is little information.

When analyzing the web applications by headquarter of the web application, it was obtained that only 38 applications provided this information through the tools used, which shows that Spain and the United States concentrate most of the headquarters of these applications.

AVERAGE VISIT DURATION (minutes): The average visit time was analyzed, resulting in three minutes for the group of study applications, the minimum is 44 seconds and corresponds to the IRRIQULTURE application and the maximum is 18 minutes with the SYNCHROTEAM application based in France.

BOUNCE RATE: The bounce indicator of this group of applications shows an average of 64%. The minimum of the indicators corresponds to the SYNCHROTEAM application of the pest phase, showing to be one of the best applications and with greater permanence on the site. The maximum of the indicator corresponds to the GESTINPOLIS application of the crop phase with an indicator of 92.30% and an average permanence at the site of 1 minute and 16 seconds. Of the 114 applications under study, only 45 have this indicator. The distribution is shown below, from which it can be seen that 25% of the applications considered in this metric have an indicator lower than 46%, that is, 11 applications have a good bounce rate indicator, since in this case it is better for the indicator to be smaller.

5 Conclusion Future Research

Given the impetus that the development of applications for agricultural activities has had, it is advisable to evaluate some of them and use them in the production of the crop. Some have a cost, which should be considered as an investment and not an expense. For the most part, their language is English because they are foreign tools.

The applications help farmers in the everyday decision-making process, allowing them to be more efficient and effective in all the processes involved in crops

The reviewed studies indicate positive developments that include an increasing number of both mobile and web applications that have developed rigorous methodologies for the collection and analysis of data, with the welcome contribution of developing countries and research institutes. Most applications are easily accessible as long as users have access to basic smart phones, because most of the applications reviewed interact with sensors, cameras and GPS. And presumably the sensors are available in almost all cases.

As future work the development of mobile applications has been considered. They may facilitate the commercialization of products, especially for small and medium producers who often have difficulties to sell, so they end up with economic losses. In this way, the production cycle could be closed with the final stage of commercialization of those products.

References

1. Food and Agriculture Organization for the United Nations: FAO Statistical Yearbook 2013: World food and agriculture. Roma (2013). www.fao.org/publications
2. República del Ecuador: Plan Nacional de Desarrollo 2017-2021 Toda una Vida. Senplades. 1–148 (2017)
3. Cáceres, R., Pol, E., Narváez, L., Puerta, A., Marfà, O.: Web app for real-time monitoring of the performance of constructed wetlands treating horticultural leachates. Agric. Water Manag. **183**, 177–185 (2017)
4. Rose, R., Rose, D.C.: Decision support tools for agriculture: towards effective design and delivery (2016)
5. Montoya et al., F.G.: A monitoring system for intensive agriculture based on mesh networks and the android system. Comput. Electron. Agric. **99**, 14–20 (2013)
6. Serrano, N., Hernantes, J., Gallardo, G.: Mobile Web Apps. IEEE Softw. **30**, 22–27 (2013)
7. Tranfield, D.: Procedures for performing systematic reviews. Br. J. Manag. **14**, 207–222 (2003)
8. Brann, D., Specialist, E.G., Sciences, S.E., Tech, V.: A Comprehensive Approach Precision Farming : Management is the KEY. Virginia Cooperative Extension, Virginia (2009)

9. Monroy, D.F.G., Hernández, Á.M., Villegas, L.M.: Mobile computing system to support the management of the seed production process in crop genebanks. In: 2014 9th Computing Colombian Conference 9CCC, pp. 109–114 (2014)
10. Kumar, A., Pathak, R.K., Gupta, S.M., Gaur, V.S., Pandey, D.: Systems biology for smart crops and agricultural innovation: filling the gaps between genotype and phenotype for complex traits linked with robust agricultural productivity and sustainability. Omics J. Integr. Biol. **19**, 581–601 (2015)
11. Bueno-Delgado, M.V., Molina-Martínez, J.M., Correoso-Campillo, R., Pavón-Mariño, P.: Ecofert: an android application for the optimization of fertilizer cost in fertigation. Comput. Electron. Agric. **121**, 32–42 (2016)
12. Suprem, A., Mahalik, N., Kim, K.: A review on application of technology systems, standards and interfaces for agriculture and food sector. Comput. Stand. Interfaces **35**, 355–364 (2013)
13. Tan, L.: Cloud-based decision support and automation for precision agriculture in orchards. IFAC-PapersOnLine **49**, 330–335 (2016)
14. Opara, L.U., Vol, E., Opara, L.U., Vol, E.: Traceability in agriculture and food supply chain : a review of basic concepts, technological implications, and future prospects. Food Agric. Environ. 1, 101–106 (2003)
15. Martens, D.C., Westermann, D.T.: Fertilizer application for correcting micronutrient deficiencies (1991)
16. Kaur, S., Dhindsa, K.S.: Comparative study of android-based M-apps for farmers. In: BT - International Conference on Intelligent Computing and Applications. Presented at the (2018)
17. Fairhurst, T.: Fertilizer chooser-an app for iOS and android (2018)
18. Mahajan, G., Prajapati, V., Singh, N.: Fertilizer calculator Goa: an android app (2015)
19. Marin, J., Reimche, C., Arciga, O., Guzman, J.C., Arciga, J., Soria, D.: Nutrienttechnologies (2018). https://nutrienttechnologies.com/about-us/
20. Hefty, D., Hefty, B.: Ag Ph.D Crop Nutrient. http://www.agphd.com/resources/ag-phd-mobile-apps/ag-phd-crop-nutrient-deficiencies/
21. Hefty, D., Hefty, B.: Fertilizer Removal. http://www.agphd.com/resources/ag-phd-mobile-apps/ag-phd-nutrient removal by crop-app/
22. PLM MEXICO SOCIEDAD ANONIMA DE CAPITAL VARIABLE: PLM Agroquímicos
23. Hefty, D., Hefty, B.: Ag PhD Field Guide. http://www.agphd.com/resources/ag-phdmobile-apps/ag-phd-field-guide/.%0D
24. mySoil: iSOYLscout. https://www.soyl.com/index.php/services/soyl-apps/isoylscout/445-isoylscout.%0D
25. Agrobase: Agrobase - weed, disease, insect. https://agrobaseapp.com/
26. Argoncontroldeplagas: Argon Control de Plagas. http://www.argoncontroldeplagas.com/
27. Geocampo agricultura de Precisión: GEOCAMPO. http://site.geocampo.co/
28. AgroPestAlert: Agropestalert. http://agropestalert.com/
29. Smart Farm System: SmartFarm. https://www.smartfarm.ag/
30. Department of Agriculture & Cooperation and Farmers Welfare, Ministry of Agriculture and Farmers Welfare, G. of I.: Crop Insurance Calculator
31. Beyondagronomy: Seeding rate Calculator. http://beyondagronomy.com/apps
32. University of Illinois Extension: Sprayer Calibration Calculator, https://web.extension.illinois.edu/state/apps.cfm
33. Agrile: Pocket Spray Smart. https://www.agrible.com/#mobileapps.%0D
34. Giroux, D.: Calibr Agro. https://itunes.apple.com/ec/app/calibragro/id972529363?mt=8
35. efarmer: eFarmer. https://efarmer.mobi/
36. aquariego: AQUA RIEGO
37. aquArson el riego inteligente: aquArson. http://aquarson.com
38. Inventia Agrarica S L: CULTIVAPP. https://www.cultivapp.com/v2.0.7/

39. Syngenta Agro SA de CV: Syngenta Producciones. http://agromaqtv.com/appsagro/apps-agricultura-de-precision/app-syngenta-soluciones/
40. Dimitri De Kerf: Gardroid – Premium.. https://play.google.com/store/apps/details?id=com.hookah.gardroid
41. NovaSource: NOVASOURCE. http://www.novasource.com/en
42. Global DPI Lic: Measure Map Lite. https://www.measuremapapp.com/
43. Studio Noframe: GPS Fields Area Measure. https://gps-fields-areameasure.uptodown.com/android
44. Evolo, S.A.: Evolo. http://evolo.online/servicios/sig/
45. Kimura, H.: Why app store keyword rankings drop dramatically seven days after launch
46. Park, S., Kang, J.: Using rule ontology in repeated rule acquisition from similar web sites. IEEE Trans. Knowl. Data Eng. **24**, 1106–1119 (2012)
47. Kumar, A.A., Jaison, J., Prabakaran, K., Escobar, J.H.: Comparative case studies on Indonesian higher education rankings comparative case studies on indonesian higher education rankings (2018)
48. Singal, H., Kohli, S.: Trust necessitated through metrics : estimating the trustworthiness of websites. Proc. – Proc. Comput. Sci. **85**, 133–140 (2016)

Author Index

R. Valencia-García et al. (Eds.): CITAMA 2019, AISC 901, pp. 143–144, 2019.
https://doi.org/10.1007/978-3-030-10728-4

Printed in the United States
By Bookmasters